PRINCIPLES AND PRACTICE OF ELECTRICAL ENGINEERING EXAMINATION P&P/EE

James W. Morrison

ARCO PUBLISHING COMPANY INC.
219 Park Avenue South, New York, N.Y. 10003

Published by Arco Publishing Company, Inc.
219 Park Avenue South, New York, N.Y. 10003

Copyright © 1977 by James W. Morrison

All rights reserved. No part of this book may be reproduced, in any form or by any means, without permission in writing from the publisher, except by a reviewer.

Library of Congress Cataloging in Publication Data

Morrison, James Warner, 1940-
   Principles and practice of electrical engineering examination.

   1. Electric engineering—Examinations, questions, etc.   I. Title.

TK169.M64   621.3'076   76-44224
ISBN 0-668-04031-9 (Paper Edition)

Printed in the United State of America

PREFACE

If you are planning to take the PRINCIPLES AND PRACTICE OF ENGINEERING EXAMINATION (P&P), this book is indispensable for a higher score.

You are well aware that the P&P is one of the most important examinations which you will ever take in your engineering career. The results of this test will determine whether you will become a registered engineer. Your entire future may well depend on a passing score on the P&P examination.

There will be many other candiates taking the NATIONAL ELECTRICAL ENGINEERING EXAMINATION (NCEE), and not all will score well enough to obtain a passing grade as established by the state board of engineer registration, and thereby, be denied professional registration status.

The PRINCIPLES AND PRACTICE OF ELECTRICAL ENGINEERING EXAMINATION(P&P), determines if a candidate meets the requirements for professional engineer registration established by state law. This book is a review of the problems covered by the national Electrical Engineering Examination; all subjects are those normally associated with the requirements for the profession in engineering education.

This book is designed to do just what the title says -- to guide you in your study so that you will have a PASSING SCORE ON THE PRINCIPLES AND PRACTICE OF ENGINEERING EXAMINATION. This book is designed specifically to aid you in preparing for the national P&P Examination.

The questions used in the practice examinations are based on actual, recent examinations, and are representative of the types of problems one may expect on future examination. The problems, with detailed solutions, have been written to assist practicing engineers, engineering students, and others, in preparing for the P&P Examination given by individual states. The examination guide has been designed specifically to simulate the uniform P&P Examination. The actual practice examination problems are similar in type, and level of difficulty; the morning and afternoon sessions are the same format; and, the practice P&P Examination can be taken as an actual 8-hour test. The book, in addition, contains a review guide of electrical engineering subjects and how to apply for

4 / PREFACE

professional engineer registration.

The problems and answers in this book have been checked several times for correctness; candidates should realize that engineering solutions often are open to different interpretations and other errors. In all cases, the State Board of Registered Professional Engineering in each state determines what is acceptable.

Part I, the introduction of the book, has information that will familiarize you with the actual morning and afternoon sessions of the PROFESSIONAL ELECTRICAL ENGINEERING EXAMINATION.

Part II offers completely solved and explained problems normally associated with the requirements for the professional electrical engineering education, and contains simulated ELECTRICAL ENGINEERING EXAMINATION questions and answers that are virtually the same as those on recent actual P&P examinations, and are representative of the types of problems you may expect on future examinations. These problems are the same level of difficulty as the ones on P&P Examinations.

How and where to apply for the P&P Examination. Discussion of state board of engineer registration procedures, including the routes and steps to professional engineering registration. How to apply to registration boards for professional status. Addresses of the state board of engineer registration are listed.

This book can be used as an independent study guide or as a text supplement for P&P Exam Review courses. The practice P&P questions and solutions will provide a valuable refresher. The reader should also study and review other texts while preparing for the preliminary examination.

In the organization and preparation of the material presented in this book, the editor has been assisted by many individuals. The editor expresses his thanks and appreciation to the many State Board of Registered Professional Engineers for their assistance in providing registration information and related P&P materials.

Thanks are extended to Dr. Robert Faiman, P.E., for the professional electrical engineering regristration material; Prof. Ramon J. Martinez, P.E., for practice electrical engineering examinations; Dr. Donald Melvin, P.E., for the illustrated circuits problems; and with the assistance of Robert Bollas, M.S., for the illustrated problems in controls, communications, and electronics; Dr. Joseph B. Murdoch, P.E., for the illumination; and Prof. Eugene A. Fucci, P.E., for the engineering economy and review of the fundamentals of electrical generators and motors.

James W. Morrison

Contents

Part I: ELECTRICAL ENGINEERING — 7

  1. Introduction to the Electrical Engineering Profession — 9
  2. Professional Engineering Registration — 15
  3. Principles and Practice of Electrical Engineering Examination — 19
  4. First Sample Principles and Practice of Electrical Engineering Examination — 25
      Solutions — 33

Part II: ILLUSTRATED PROBLEMS — 41

  5. Circuits — 43
  6. Controls — 69
  7. Communications — 89
  8. Electronics — 103
  9. Illumination — 127
  10. Engineering Economy — 147

Part III: SECOND SAMPLE PRINCIPLES AND PRACTICE OF ELECTRICAL ENGINEERING EXAMINATION — 175

  Morning Section Questions — 177
      Solutions — 183
  Afternoon Section Questions — 195
      Solutions — 200

Part IV: THIRD SAMPLE PRINCIPLES AND PRACTICE OF ELECTRICAL ENGINEERING EXAMINATION — 211

  Morning Section Questions — 213
      Solutions — 218
  Afternoon Section Questions — 231
      Solutions — 237

Part V: APPENDICES 247
   1. Review: Fundamentals of Electrical Generators
       and Motors 249
   2. Engineering Registration 267
   3. Addresses of State Registration Boards 289
   4. Test-Taking Made Simple 297

Part I

Electrical Engineering

Chapter 1

INTRODUCTION TO THE ELECTRICAL ENGINEERING PROFESSION

The work of engineers affects our lives in thousands of different ways. Their past accomplishments have enabled us to drive safer automobiles, reach the moon, and even prolong life through special machinery. Future accomplishments could help us obtain energy self-sufficiency, develop more pollution-free powerplants and aid medical science's fight against disease.

In 1975 more than 1.2 million persons were employed as engineers, the second largest professional occupation exceeded only by teachers. The number of women engineers (about one percent) is expected to increase in the future since enrollments of women in engineering programs have increased sharply over the past several years.

Electrical engineers work in medicine, computers, missile guidance, or electric power distribution. Because engineering problems are usually complex, the work in some fields cuts across the traditional branches. Using a team approach to solve problems, engineers in one field often work closely with specialists in other scientific, engineering, and business occupations.

Electrical engineers design, develop, and supervise the manufacture of electrical and electronic equipment. These include electric motors and generators; communications equipment; electronic equipment such as heart pacemakers, pollution measuring instrumentation, radar, computers, lasers, and missile guidance systems; and electrical appliances of all kinds. They also design and operate facilities for generating and distributing electric power.

Electrical engineers generally specialize in a major area of work such as electronics, computers, electrical equipment manufacturing, communications, or power. Others specialize in subdivisions of these broad areas like microwaves or missile guidance and tracking systems. Many are engaged in research, development, and design activities. Some are in administrative and management jobs; others work in manufacturing operations, in technical sales, or in college teaching.

Electrical engineering is the largest branch of the profession. Over 300,000 electrical engineers were employed in 1975, mainly by manufacturers of electrical and electronic equipment, aircraft and parts, business machines, and professional and scientific equipment. Many work for telephone, telegraph, and electric light and power companies. Large numbers are employed by government agencies and by colleges and universities. Others work for construction firms, for engineering consultants, or as independent consulting engineers.

Training, Other Qualifications, and Advancement

A bachelor's degree in engineering is the generally accepted educational requirement for beginning engineering jobs. College graduates trained in one of the natural sciences or mathematics also may qualify for some beginning jobs. Experienced technicians with some engineering education are sometimes able to advance to engineering jobs.

Graduate training is being emphasized for an increasing number of jobs; it is essential for most beginning teaching and research positions, and desirable for advancement. Some specialties, such as nuclear engineering, are taught mainly at the graduate level.

About 280 colleges and universities offer a bachelor's degree in engineering. Although programs in the larger branches of engineering are offered in most of these institutions, some small specialties are taught in only a very few. Therefore, students desiring specialized training should investigate curriculums before selecting a college. Admissions requirements for undergraduate engineering schools usually include high school courses in advanced mathematics and the physical sciences.

In a typical 4-year curriculum the first 2 years are spent studying basic sciences -- mathematics, physics, chemistry, introductory engineering -- and the humanities, social sciences, and English. The last 2 years are devoted, for the most part, to specialized engineering courses. Some programs offer a general engineering curriculum permitting the student to choose a specialty in graduate school or acquire it on the job.

Some engineering curriculums require more than 4 years to complete. A number of colleges and universities now offer 5-year master's degree programs. In addition, several engineering schools have formal arrangements with liberal arts colleges whereby a student spends 3 years in liberal arts and 2 years in engineering and receives a bachelor's degree from each.

Some schools have 5- or even 6-year cooperative plans where students coordinate classroom study and practical work experience. In addition to gaining useful experience, students can finance part of their education. Because of the need to keep up with rapid advances in technology, engineers often continue their education throughout their careers in programs sponsored by employers, or in colleges and universities after working hours.

All 50 States and the District of Columbia require licensing for engineers whose work may affect life, health, or property, or who offer their services to the public. In 1974, about 350,000 engineers were registered. Generally, registration requirements include a degree from an accredited engineering school, 4 years of relevant work experience, and the passing of State examinations, e.g., Engineer-In-Training (EIT) and/or Principles and Practice of Electrical Engineering (P&P-LE).

Engineering graduates usually begin work under the supervision of experienced engineers. Many companies have special programs to acquaint new engineers with special industrial practices and to determine the specialties for which they are best suited. Experienced engineers may advance to positions of greater responsibility; those with proven ability often become administrators and increasingly larger numbers are being promoted to top executive jobs. Some engineers obtain graduate degrees in business administration to improve their advancement opportunities, while still others obtain law degrees and become patent attorneys.

Engineers should be able to work as part of a team and have creativity, an analytical mind, and a capacity for detail. They should be able to express their ideas well orally and in writing.

Employment Outlook for Electrical Engineers

Employment opportunities for engineers are expected to be good through the mid-1980's. Opportunities for recent graduates of engineering schools are expected to be very good since the number of new graduates is expected to fall short of the number needed to fill the thousands of openings created by employment growth, and the need to replace those who die, retire, or transfer to other occupations. Because of the expected shortage, many openings will be filled by upgraded technicians and college graduates from related fields.

Employment requirements for engineers are expected to grow faster than the average for all occupations through the mid-1980's. Much of this growth will stem from industrial expansion to meet the demand for more goods and services. More engineers will be needed in the design and construction of factories, electric powerplants, office buildings, and transportation systems, as well as in the development and manufacture of more advanced computers, scientific instruments, industrial machinery, chemical products, and motor vehicles.

Many engineers will be required in energy-related activities developing new sources of energy as well as designing energy-saving systems for automobiles, homes, and other buildings. Engineers also will be needed to solve environmental pollution problems.

Defense spending will also affect the outlook for engineers since a large number work in defense-related activities. The long-range outlook for engineers given here is based on the assumption that defense spending will increase from its 1975 level by the mid-1980's, but will still be somewhat lower than the peak levels of the 1960's. If, however, defense activity differs substantially from the level assumed, the demand for engineers will differ from that now expected.

Employment of electrical engineers is expected to increase faster than the average for all occupations through the mid-1980's. Increased demands for products such as computers, communications, and electric power generating equipment, and military electronics is expected to be the major factor contributing to this growth. The demand for electrical and electronic consumer goods, along with increased research and development in nuclear power generation, should create additional jobs for electrical engineers. Many electrical engineers also will be needed to replace personnel who retire, die, or transfer to other fields of work.

The long-range employment for electrical engineers is based on the assumption that defense spending in the mid-1980's will increase from the 1975 level, but will still be somewhat lower than the peak level of the late 1960's. If defense activity should differ substantially from the projected level, the demand for electrical engineers will differ from that now expected.

Employment in electronics manufacturing is expected to increase faster than the average for industries through the mid-1980's. In addition to the jobs resulting from employment growth, large numbers of openings will arise as experienced workers retire, die, or take jobs in other industries.

Although employment in the electronics industry is expected to grow over the long run, it does fluctuate from year to year becuase of changes in economic activity and defense spending. As a result, job openings may be plentiful in some years, scarce in others.

Employment of electrical engineers is expected to increase faster than total employment because of continued high expenditures for research and development and the manufacture of more complex products. Among professional and technical workers, the greatest demand will be for engineers, particularly those who have a background in certain specialized fields, such as quantum mechanics, solid-state circuitry, product design, and industrial engineering. Many opportunities also will be available for engineers in sales departments because the industry's products will require sales personnel with high technical backgrounds. The demand for mathematicians and physicists will be particularly good because of expanding research in computer and laser technology.

Sources of Information

General information on engineering careers including guidance, professional training, and salaries is available from:

ELECTRICAL ENGINEERING PROFESSION / 13

Engineers' Council for Professional Development, 345 East 47th Street, New York, New York 10017.

Engineering Manpower Commission, Engineers Joint Council, 345 East 47th Street, New York, New York 10017.

National Society of Professional Engineers, 2029 K Street, N.W., Washington, D.C. 20006.

Information on registration of engineers may be obtained from:

National Council of Engineering Examiners, P.O. Box 752, Clenson, S.C. 29613.

For information about graduate study, contact:

American Society for Engineering Education, One Dupont Circle, Suite 400, Washington, D.C. 20036.

For professional engineering organization information:

Aerospace Electrical Society (AES), P.O. Box 24BB3, Village Station, Los Angeles, California 90024.

American Federation of Information Processing Societies, Incorporated (AFIPS), 210 Summit Avenue, Montvale, N.J. 07645.

American Society for Engineering Education (ASEE), National Center for Higher Education, One Dupont Circle, Suite 400, Washington, D.C. 20036.

Association for Computing Machinery (ACM), 1133 Avenue of the Americas, New York, New York 10036.

Illuminating Engineering Society (IES), 345 East 47th Street, New York, New York 10017.

The Institute of Electrical & Electronics Engineers (IEEE), 345 East 47th Street, New York, New York 10017.

Insulated Power Cable Engineers Association (IPCEA), 192 Washington Street, Belmont, Massachusetts 02178.

The International Commission on Illumination, U.S. National Committee (USNC/CIE), c/o National Bureau of Standards, Washington, D.C. 20234.

International Conference on Large High Tension Electric Systems, U.S. National Committee (CIGRE), c/o Ebasco Services, Inc., Two Rector Street, New York, N.Y. 10006.

14 / ELECTRICAL ENGINEERING PROFESSION

>National Association of Government Engineers (NAOGE), 815 15th Street, N.W., Room 711, Washington, D.C. 20005.
>
>National Association of Power Engineers (NAPE), 176 West Adams Street, Suite 1914, Chicago, Illinois 60603.
>
>National Electronics Conference, Incorporated (NEC), 1301 West 22nd Street, Oak Brook, Illinois 60521.
>
>National Society of Professional Engineers (NSPE), 2029 K Street, N.W., Washington, D.C. 20006.
>
>Radio Technical Commission for Aeronautics (RTCA), Suite 655, 1717 H. Street, N.W., Washington, D.C. 20006.
>
>The Society of American Military Engineers (SAME), Union Trust Building, 705 14th Street, N.W., Washington, D.C. 20005.

Many other engineering organizations are listed in the following publications available in most libraries or from the publisher:

>Directory of Engineering Societies, published by Engineers Joint Council, 345 East 47th Street, New York, N.Y. 10017.
>
>Scientific and Technical Societies of the United States and Canada, published by the National Academy of Sciences, National Research Council.

Some engineers are members of labor unions. Information on engineering unions is available from:

>International Federation of Professional and Technical Engineers, 1126 16th Street, N.W., Washington, D.C. 20036.

Chapter 2

PROFESSIONAL ENGINEERING REGISTRATION

In 1729 B.C., Hammurabi, the sixth of eleven kings in the Old Babylonian (Amorite) Dynasty, promulgated his famous lawcode which provided protection of the uninformed layman. The Code of Laws of Hammurabi contained the following law of the land or regulations dealing with builders:

> If a builder constructed a building, the owner of the new structure shall pay appropriately for the building as proper remuneration.
>
> If a builder constructed a building but did not make his work strong, with the result that the structure collapsed and so caused the death of the owner of the building, the builder shall be put to death.
>
> If it has caused the death of a son of the owner of the structure, they shall put the son of that builder to death.
>
> If it caused the death of a slave of the owner of the structure, he shall give a slave for slave to the owner of the building.
>
> If it has destroyed goods, he shall make good whatever it destroyed; also because he did not make the structure strong which he built and it collapsed, he shall reconstruct the building which collapsed at his own expense.
>
> If a builder constructed a structure for the owner and has not done his work properly so that a wall has become unsafe, that builder shall strengthen that wall at his own expense.

These were severe penalties for failure of a builder to do work properly; however, no proof of competence was required before the work was undertaken.

Licensing of engineers was late in coming to the United States; in 1965 the first National Council of Engineering Examiners uniform examination was offered to the state registration boards. Thirty boards used this examination and the central grading service. By April 1974, only four of the fifty-five Member Boards of National Council did not use any of the examination material relating to engineering.

The Boards of Engineering Examiners in the various states, District of Columbia, Puerto Rico, Guam, and the Canal Zone administer the NCEE's Engineer-in-Training (EIT) Examination and the Principles and Practices (P&P) Examination. These examinations are designed to demonstrate the adequacy of the equivalent knowledge and understanding and the Boards will issue a certificate to those who pass either examination.

Registration acts, state board rules and regulations require that an applicant must demonstrate qualifications in a major field of engineering. In addition to a formal education and experience at the professional level under qualified engineering supervision, the applicant must show the ability to apply sound engineering principles and judgment to the solution of problems normally encountered in practice. Because of the wide variety of work in any one major field, registration as a Professional Engineer should indicate that one has a reasonable working knowledge in more than one narrow specialty. To establish the qualification of a given applicant in a one-day examination, a more reliable test is given through the use of several relatively short problems rather than a few very long and involved ones.

Registration as a "professional engineer" through a state board of registration is the only legal basis for the public practice of engineering. All state statutes and laws currently exempt engineers employed by industry or the Federal government; efforts are under consideration in some states to eliminate at least the industry exemption. The basis for engineering registration (licensing) is the obligation of the state for protection of the safety, health, and welfare of the public. Each state, territory, and the District of Columbia has a formal statute under which engineers are examined and registered; implementation of the mechanics of the registration process is carried out under "boards of registration." Registration to practice in any state must be obtained from that state; approval of an applicant registered in another state (comity) is based on satisfaction of the minimum standards of each board to which application is made. Each state is autonomous and therefore state statutes and board regulations are not uniform.

The practice of engineering is defined as based on an application of the basic principles of mathematics, physical sciences and the engineering sciences. Boards of registration evaluate education and experience as a basis for registration. While engineering is one of the slowest of the professions to formally require minimum educational requirements some states have or will shortly require an accredited engineering degree as the minimum standard for consideration and in almost all states registration is facilitated by the holding of a degree. Experience "of a nature satisfactory to the board" is a requirement by all boards. While some variation exists between boards in the amount of experience required, it in all cases must be of a professional nature.

While each board of registration is autonomous, there is a general pattern to their procedures even though details and specific requirements may vary. The evaluation of a candidate by a board involves a judgment of education and experience. The judgment process usually includes formal examination in 1) fundamentals of engineering (often referred to as the Engineer-in-Training -- EIT -- examination) and 2) Principles and Practice of Engineering. Both examinations are eight hours in length and most states utilize common instruments prepared by the National Council of Engineering Examiners (a co-operative body established by the state boards to facilitate their operations); the NCEE also provides a "certification" service for engineers requiring registration in a number of states which enables state boards to simplify their evaluation processes. Educational and experiental requirements for admission to examination vary and each individual must consult the board of registration in the state of residence for specific information.

Chapter 3
PRINCIPLES AND PRACTICE OF ELECTRICAL ENGINEERING EXAMINATION

The NCEE's Principles and Practice Examination is usually taken as part of the process involved in achieving registration as a professional engineer. It is an 8-hour examination which provides an in-depth assessment of a candidate's knowledge and skill in a specific engineering field. This examination normally is taken after successful completion of the EIT examination, although a few states allow a candidate to take the exam based on experience only. Since the P&P exam is of a narrower and more specialized nature, less emphasis on fundamentals may be expected. However, the candidate must be sure that his experience is sufficiently broad to cover the whole of the speciality chosen.

Since the principal purpose of any examination is to determine the competency of the individual taking the examination written examinations are given candidates for professional engineering registration to determine if they meet the minimum requirements for registration established by law. These examinations should ascertain: (1) if the candidate has an adequate understanding of the basic and engineering sciences, and (2) if his training and experience have taught him to apply these basic and engineering sciences to the solution of engineering problems.

The NCEE Examination Committee has suggested that the electrical engineering field be divided into categories. Furthermore, the Committee has recommended the number of problems that should be given in each of these categories. The list below shows the major categories along with the number of problems recommended for each. However, there is no assurance that precisely this number of problems will be available in the respective categories for any given eight hour examination. This list should serve only as a guide.

Major field of Engineering	Category	Subject	Approximate Number of Problems per Category
Electrical*	A	Illumination	2
	B	Machines	2
	C	Electronics	2

D	Communication	2
E	Circuits	3
F	Controls	2
G	Engineering Economy	2
H	Instrumentation	2
I	Power & Systems	3

* There could also be another category which would include all major fields of engineering or another electrical engineering topic e.g., digital logic.

Questions for the engineering examinations come from state boards, colleges, industry, participating organizations and other sources. Under procedures used as this is prepared, the NCEE executive staff prepares proof copies of the forthcoming examinations for review by the Uniform Examinations Committee. Upon approval by the committee, the examinations are printed and distributed to the Member Boards for administration on a set schedule twice a year in April and November. The Uniform Examinations Committee and the NCEE executive staff work jointly in research and development to ensure continued improvement and timeliness of the examinations.

The questions given in this book were taken from recent examinations between 1971-1976. These are believed to be representative of the type questions that one might expect on a national examination. NCEE has a policy of denying access to others in publishing the examination problems, therefore, none of the problems in this publication are from the P&P examination. The problems in this book do reflect professional evaluation of the kinds and distribution of problems which have appeared on the recent (actual) NCEE examinations in electrical engineering. The applicant is expected to select and work four from a collection of ten such problems in a four hour period. Questions on Economic Analysis are common to all fields of engineering and these are included on all examinations. However, there will not be more than one problem on economics in each four hour examinations.

While the most general engineering principles are required for the solution of many of the problems as they are now given, in the future, however, the NCEE may not limit the range of questions to that now used. An attempt is made to keep pace with the changing engineering profession. The application of good engineering judgment in the selection and evaluation of pertinent information and the demonstration of an ability to make reasonable assumptions when necessary are considered to be very important factors. Answers submitted for some problems may vary according to assumptions. Errors may occur under pressure of time limitation. Partial credit is always given if correct fundamental engineering principles are used, even though the answer may be incorrect.

P & P/EE EXAMINATION / 21

In this book an attempt has been made to organize the most recent national electrical engineering problems within the usual meanings, interpretations and actual electrical engineering practice. As the electrical engineering profession advances, however, changes in the format, content and, possibly subject matter of NCEE examinations will follow suit. Efforts by the National Council are being directed constantly to make the P&P examination problems relate more directly with the "practical" rather than the "textbook".

The above is especially important when the candidate considers the 8-hour P&P examination with each session (morning and afternoon) having 10 questions from which only four questions are to be answered. Two of these eight questions can be engineering economy. Each question is valued at ten points with a maximum score of forty points per session.

Illumination

Illustrated problems as well as practice examination problems deal with lighting design and illumination requirements. P&P problems usually deal with the following: Power requirements using fluorescent lamps, mercury vapor vs. incandescent lighting, mercury vapor calculations utilizing "isolux curves", plant illumination calculations for fluorescent lamps, and a computed footcandle level of a particular layout.

Machines and Motors

This traditional field includes a broad range of questions which can be transformer connections, such as 3-phase zig-zag voltage ratios and loading; transformer efficiencies requiring the use of equivalent circuits from open circuit; complete dc motor analysis involving starting torques and efficiency calculations; ac machines and loading. These are usually problems of 3-phase induction motor analysis, induction motor equivalent circuit analysis, the use of inverter to drive motors, and the design of dc motor starter circuit, equivalent circuit calculation, or dc shunt motor analysis.

Electronics

Some typical questions include the following: push-pull vacuum tube power stage; 2-coupled transistor stages; 3-stage direct coupled transistor amplifier; differential amplifier using transistors; and 2-stage FET, transistor amplifier. Additional areas of P&P questions involve the use of transistor logic circuit and truth table for digital logic problems, transistor amplifier bandwith calculation and tuned transformer-coupled transistor amplified.

Communication

P&P problems involve transmission lines, use of the Smith chart, field and waves and full communication systems problems. The following is a

listing of typical questions.
> Match an antenna to a line by Smith Chart.
> Full communication system using AM, FM, modulation.
> Frequency response of a filter to a time signal.
> Digital transmission system for voice signal.

Circuits

Usually at least one or more problems involves an RLC transient or an RLC series parallel combination circuit using steady ac circuit analysis.

Another frequently asked circuits problem is one involving filter theory, either constant "K" or "M" derived.

The following is a list of circuit problems included: ac circuits, RLC transient, filter network calculation, RLC-differential equation to Laplace Transformation, matrix form of network equation, series RLC resonance and "bandwidth" analysis, steady state RLC ac circuit analysis.

Controls

Many of the NCEE problems ask something about system stability for a linear control system. Other question include: steady-state error and specifications for system, lead network compensation design, root locus plot, and stability analysis for unity feedback system.

Engineering Economy

Compute uniform annual amount equivalent to an irregular series of money amounts.

Two alternatives, one with perpetual life. Fixed output. Compute uniform equivalent annual cost.

Present worth of a bond.

Two alternative rent or buy equipment questions.

Unit cost for various production lot sizes.

Optimum allocation of load to minimize generating plant fuel cost.

Breakeven analysis by method of equivalent annual cost.

Two alternatives, compute expected present worth of cost.

Instrumentation

Instrumentation and measurement questions are usually a combination of the above fields and involve a specific application problem, such as, voltage regulation using nonlinear resistance, field winding protection using diodes, wave shaping circuit using diodes, wave form analysis from diode clipped sinusoids, power output from SCR circuit, power supply regulation using zener diodes, steady state phase shifting network analysis, or temperature measurment with balancing bridge circuit.

Power and Systems

Usually one problem involves power factor correction either by capactors or using a synchronous motor. Other problems include: total power calculation from various power factor loads, 3-phase transmission line calculation, and open delta transformer loading.

Basic Theory

A review unit of basic electrical theory has been included:

Electric Circuits, Transformers, DC Manchines, and Motors

Review AC circuits - phasors, complex impendances and admittances, series and parallel circuits.
Power and power factor correction - average power, reactive power, power factor, power factor correction.
Coupled circuits - mutual inductance, coupling co-efficient, dot convention, analysis of coupled circuits.
Polyphase circuits - wye and delta connected sources, line-to-line and phase voltages, line and phase currents, balanced wye and delta loads, three phase power.
Magnetic circuits - magnetic flux, magnetic field intensity, ferromagnetic materials, magnetic circuit calculations.
Transformers - operation, eddy currents, hystersis, equivalent circuit, efficiency, regulation.
DC machines - electromotive force in elementary machine, basic voltage and current relations, equivalent circuit, series, shunt and compound machines.
Alternator - voltage, torque, frequency, poles, circuit model, phasor diagram.
Synchronous motor - principle of operation, excitation, phasor diagram.
Induction motor - princple of operation, circuit model, speed-torque characteristics, slip.
Single phase induction motors - principles of operation.

Mesh Current, Thevenin's Norton's and Network Theorems

Mesh Current Network Analysis; Choice of Mesh Currents; Number of Mesh Currents Required; Mesh Equations.
Thevenin's Theorem.
Norton's Theorem.
Thevenin and Norton Equivalent Circuits.
Network Theorems; Wye-delta transformation; Superpostion Theorem.

While many electrical engineers see the nature and scope of the NCEE's professional electrical engineering examination somewhat beyond their own day-to-day job responsibilities, there is a recognition of the examination structure and flexibility, i.e., only one third of the available problems must be answered. There is also the realization by most electrical engineers that a substantial technical review program, prior to the electrical engineering principles and practice examination is a requirement. This book is an attempt by several registered electrical engineers to assist others in organizing such a review program.

Chapter 4

FIRST SAMPLE PRINCIPLES AND PRACTICE OF ELECTRICAL ENGINEERING EXAMINATION

You will have four hours in which to work this test. Your score will be directly proportional to the number of problems you solve correctly through four (4). Each correct solutions counts ten points. The maximum possible score for this part of the examination is 40 points. Partial credit for partially correct solutions will be given.

Work four of the problems according to instructions. Do not submit solutions or partial solutions for more than four problems. Indicate the problems which you have solved.

You may work only one engineering economy problem. When you have completed this portion of the P&P examination in the required time limit, you should check your solutions with the answers at the end of this examination. For additional practice you are encouraged to work the other problems under the time limit.

Since you want the maximum points available, you should remember that the examiner who assigns these points must make his judgment based only on what has been written down during the examination. It is important for you to be reasonably neat in your work and write down any assumption that you consider necessary to allow you to solve the problem properly and to provide sufficient rationale so that the examiner can judge your reasoning. Assumptions should follow the logic and requirements of the problem.

You are advised to use your time effectively.

1. A company has received bids on a 2,000 kva (kilovolt-amperes) transformer from several manufacturers. One of these quotes $15,000 for a transformer having 98% efficiency at full load, 97% at three-quarter load, and 93% at half load. Another firm quotes $13,750 for a transformer having 96% efficiency at full load, 95% at three-quarter load, and 90% at half load. The company expects to keep the transformer on this service for 15 years, at which time salvage values are expected to be $300 and $275 respectively. Aside from these differences, the transformers are essentially equal.

Energy costs 2 cents per kilowatt-hour. The company's minimum required rate of return is 15%. Annual utilization of the transformer is expected to be 600 hours at full load, 1,800 hours at three-quarter load, and 600 hours at half load. Assume that the company's load is at unity power factor so that full load on the transformer is 2,000 kw. Make a recommendation from an annual-cost comparison.

2. A constant-current regulated supply, shown below, is to provide a test current of 10 mA over the compliance range of the supply. Determine the necessary component values when the following quantities and conditions are known:

I_Z = 2 mA $\qquad$ h_{FE} = 100

E_{FL} = 40 V $\qquad$ I_{FL} = 12 mA

h_{ib} = 20 Ω $\qquad$ V_{EB} = 0.2 V

What is the compliance range of this supply?

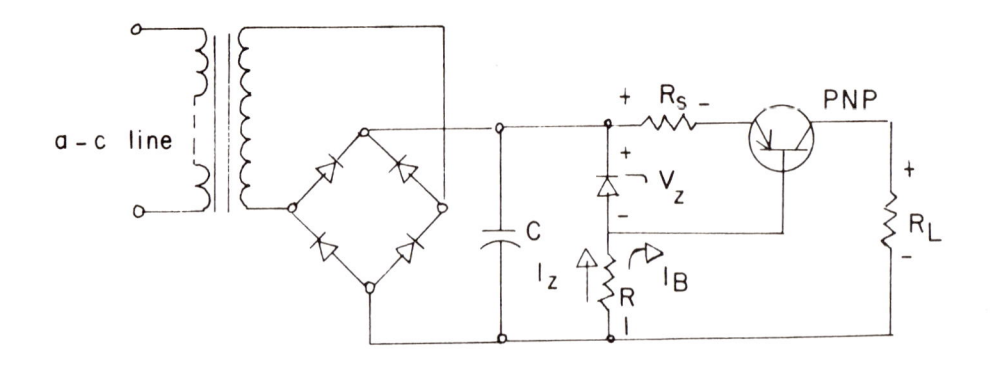

3. A 10-kw, 125-v, 1800-rpm shunt generator is operated in parallel with a 5-kw, 125-v, 1200-rpm shunt generator. The 10-kw dynamo has an 8 percent while the 5-kw dynamo has a 10 percent voltage regulation, respectively. Assuming that both dynamos are parallel at a no-load voltage of 135 v and the speed is constant with application of load, for a total load current of 100 amp delivered to a load, calculate

   a) The load current delivered by each generator

   b) The kilowatts delivered to the load

   c) The kilowatts delivered by each generator

4. The amplifier shown has the following parameters:

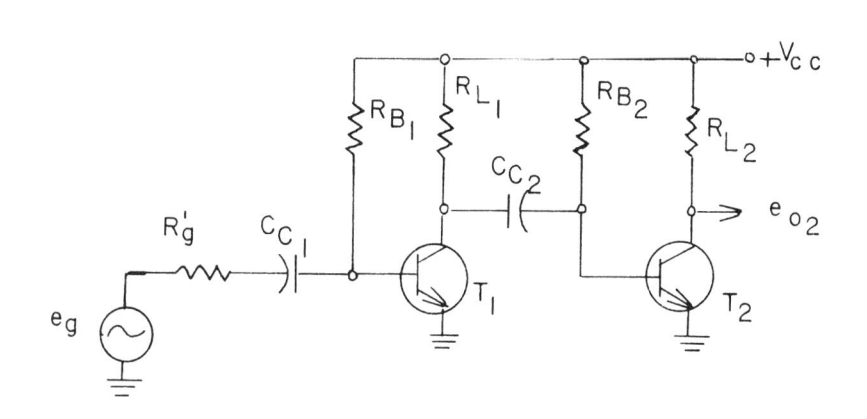

D-C Parameters

$V_{CE_{1,2}} = 5V$; $I_{C_{1,2}} = 4mA$

$V_{BE_{1,2}} = 0.2V$; $I_{B_{1,2}} = 40 \mu A$

$V_{CC} = 13V$

Signal Parameters

$h_{ie} = 1.2K\Omega$; $h_{fe} = 100$

$h_{re} = 5\times10^{-4}$; $h_{oe} = 100\times10^{-6}$

$C_{b'c} = 4$ pF; $C_{b'e} = 100$ pF

$g_m = 0.1$ mho

Find:

a) R_{B_1} and R_{B_2}

b) R_{L_1} and R_{L_2}

c) A_v and A_i overall at MF

d) C_{C_1} and C_{C_2} for a lower breakpoint-frequency of 100 H_z when $R'_g = 20K$

e) Device f_T

f) Device t_r

g) Device f_β

5. A voltage-feedback loop is to be arranged on a CE-transistor amplifier, as shown.

Determine the following:

a) R_L and R_B when $V_{CC} = 10V$, $V_{CE} = 5V$, $I_C = 2mA$, $V_{BE} = 0.2V$, $I_B = 50\ \mu A$.

b) Show the signal-flow diagram for this stage, using signal voltages as the initial and final nodes.

c) How many decibels of feedback will this amplifier have when the open-loop current gain is 40 and the open-loop voltage gains is -45?

d) What is the bandwidth extension on the high-frequency end of the response curve for the closed-loop amplifier, relative to the open-loop amplifier?

e) What is the r'_{in} of the stage?

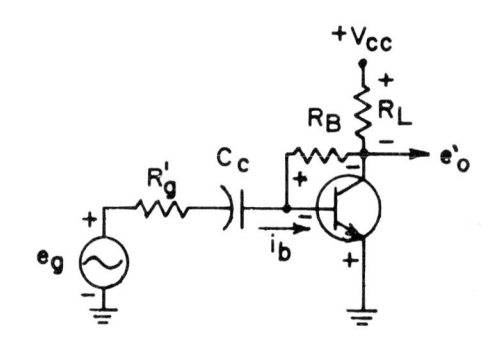

6. The interior dimensions of an industrial plant are 100 ft. width, 200 ft. length, 24 ft. height. The long axis of machines is oriented in the direction of the plant length. On illumination proposal considers using an 8-ft., 100-in. channel length luminaire housing 6-40 watt repid-start fluorescent lamps with matching 4-ft., 50-in. channel length luminairs if needed. Basic design data is as follows:

Maintained illumination level	100 ft.c.
Initial lamp lumens per lamp	3,100 lumens
Coefficient of Utilization (C.U.)	0.78
Maintenance Factor (M.F.)	0.6
Mounting height above floor	19 ft.
Wattage per 8-ft. unit including ballasts	285 watts
Wattage per 4-ft. unit including ballasts	190 watts

(1) Determine the number of 8-ft. units required.

(2) Present a sketch of the layout of luminaires for the plant indicating or specifying the luminaires per row, the number of rows, the spacing between rows, and the space between row and walls.

(3) Give number of 8-ft. units used and number of 4-ft. units used, and calculate the cost of energy per year if the average energy rate is 1.2¢ per kwh with plant operations of 17 hours per day, 5 days per week and 50 weeks per year.

7. A 400-mW semiconductor characterized as shown in the figure has a maximum junction temperature of $225°C$.

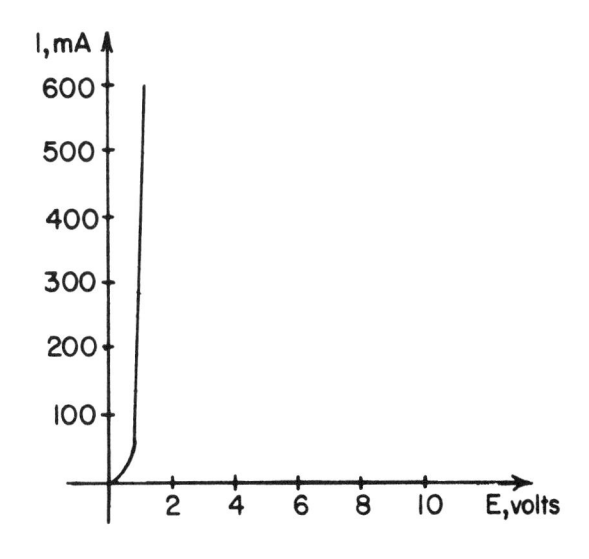

Determine:

a) The thermal resistance of the diode (the 400 -mW rating applies to $25^\circ C$ ambient operation).

b) The load resistor for I_{max} operation in a $25^\circ C$ ambient when E_{max} is 10V.

c) The load resistor for I_{max} operation in a $40^\circ C$ ambient when E_{max} is 10V. (Assume no substantial change in the forward characteristic of the diode.)

d) The ambient temperature for 300 -mW dissipation.

8. Find $3(t)$ in the circuit shown using differential equations assuming the switch is opened at time zero. The current input is an impulse.

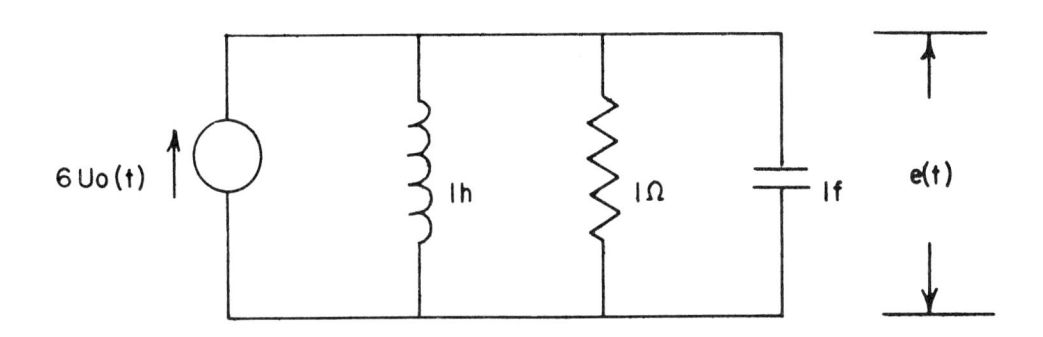

Circuit: $6 U_0(t)$ current source in parallel with 1h inductor, 1Ω resistor, and 1f capacitor; output $e(t)$.

9. The filtered output of a rectifier power supply is to furnish the voltage and current for bias of a two-stage vacuum-tube amplifier having the following requirements:

V_1:
$E_{PP} = 220V$
$I_p = 5mA$

V_2:
$E_{PP} = 260V$
$I_p = 9mA$, $I_{SG} = 3mA$
$E_{GG} = -12V$

The available supply voltage is 282V. Find the voltage-divider bleeder-resistor values and the power rating of each resistor when bleeder current is 10 percent of the load current.

Solution 1:

$$AC_A = (15,000 - 300)(crf)_{.15-.15}^{.17102} + 300(.15) + \left[\frac{2,000(600)}{.98} + \frac{1,500(1,800)}{.97} + \frac{1,000(600)}{.93}\right].02 = \$95,620$$

$$AC_B = (13,750)(crf)_{15}^{.17102} = 275(.15) + \left[\frac{2,000(600)}{.98} + \frac{1,500(1,800)}{.95} + \frac{1,000(600)}{.90}\right].02 = \$97,520$$

Take A.

Solution 2:

$R_S = 100 h_{ib} = 100 \times 20 = 2K$

$I_B = \dfrac{I_L}{h_{fe}} = \dfrac{100 ma}{100} = 100 \mu a$

$V_{RS} = (I_L + I_B) R_S = 10.1 ma \times 2K = 20.2V$

$V_Z = V_{RS} + V_{EB} = (20.2 + .2) = 20.4V$

$R = \dfrac{E_{FL} - V_Z}{I_Z + I_B} = \dfrac{40 - 20.4}{2 + .1} = 9.3K$

The compliance voltage is:

$E_{FL} - V_{RS} = 40 - 20.2 = 19.8$

which is less than the $V_{CE_{sat}}$ of the pass transistor.

Solution 3:

Gen A = 5 kw   125 v, rated load = $\dfrac{5000 \text{ w}}{125 \text{ v}}$ = 40a.

Gen B = 10 kw   125 v, rated load = $\dfrac{1000 \text{ w}}{125 \text{ v}}$ = 80a.

Solution 3 cont'd:

	Gen A	Gen B
kw	5	10
I_L amps	40	80
V.R. percent	10	8

The franction of rated current carried by each generator varies inversely as its voltage regulation when connected in parallel OR

$$\frac{\dfrac{I_A}{I_A \text{ rated}}}{\dfrac{I_B}{I_B \text{ rated}}} = \frac{V.R._B}{V.R._A}$$

then: $\dfrac{\dfrac{I_A}{40}}{\dfrac{I_A}{80}} = \dfrac{8}{10}$ and $\dfrac{2I_A}{I_B} = \dfrac{8}{10}$ ∴ $I_B = 2.5\, I_A$

a) Given $I_L = 100a$, $I_A + I_B = 100a$
$I_A + 2.5\, I_A = 100a$, $I_A = 28.6a$
$I_B = 100 - 28.6 = 71.4a$

b) $KW_{Load} = V_L I_L$

$V_{n.l.}$ for Gen B = $1.08 \times V_{fl} = 1.08 \times 125 = 135v$

V_{drop} for Gen B = $\dfrac{71.4a}{80a} \times (135-125) = 8.92v.$

$V_B = V_L = 135v - 8.92v = 126.18v$, $KW_{Load} = 126.18 \times 100a \times \dfrac{1}{1000}$

= 12.62 kw.

c) $W_A = \dfrac{126.2v \times 28.6a}{1000} = 3.61$ kw

$W_B = \dfrac{126.2v \times 71.4a}{1000} = 9.01$ kw

$W_A + W_B = 3.61 + 9.01 + 12.62$ kw

Solution 4:

a) $R_{B_{1,2}} = \dfrac{V_{CC} - V_{BE}}{I_B} = \dfrac{13 - 0.2}{40\mu a} = 320K$

b) $R_{L_{1,2}} = \dfrac{V_{CC} - V_{CE}}{I_C} = \dfrac{13 - 5}{4ma} = 2K$

c) $A_{i_{T_2}} = \dfrac{h_{fe}}{1 + h_{oe} R_L} = \dfrac{100}{1 + 100 \times 10^{-6} + 2K} = 83$

$A_i(\text{interstage 2}) = \dfrac{R_{eq_2}}{r_{in_2}} = \dfrac{750}{1.2K} = 0.625$

where $r_{in_2} \sim 1.2K$, $R_{eq_2} = R_{L_1} \| r_{in_2} = 2K \| 1.2K = 750\Omega$

$A_i(\text{interstage 1}) = \dfrac{R_{eq_1}}{r_{in_2}} = \dfrac{1.2K}{1.2K} = 1$

$A_{i_{T_1}} = \dfrac{h_{fe}}{1 + h_{oe} R_{eq_2}} = \dfrac{100}{1 + 100 \times 10^{-6} \times 750} = 93$

$A_i \text{ (overall)} = 83 \times .625 \times 1 \times 93 = 4800$

$A_v = A_i \times \dfrac{R_{L_2}}{R_{eq_1}} = 4800 \times \dfrac{2K}{1.2K} = 8000$

d) $R_a = r_{o_1} \| R_{L_1} = 10.3K \| 2K = 1.67K$

where $r_{o_1} = \dfrac{h_{ie} + R_g}{\Delta^h + h_{oe} R_g} = 10.3K$

where $R_g = R'_g \| R_{B_1} = 19K$

and $\Delta^h = .07$

$R_b = R_{B_2} \| r_{in_2} \sim r_{in_2} = 1.2K$

$C_{C_1} = \dfrac{0.159}{f_1 (R'_g + R_{eq_1})} = \dfrac{0.159}{100(19K + 1.2K)} = .08\mu f$

$C_{C_2} = 0.084\mu f$

Solution 4 cont'd

e) $f_\tau = \dfrac{g_m}{2\pi C_{b'e}} = \dfrac{0.1}{2\pi \times 100 pf} = 159 \text{Mhz}$

f) $t_r = 2.2\, h_{ie}' C_{b'e} = 2.2 \times 1K \times 100\, pf = 0.22\,\mu\text{Sec}$

where $g_m = \dfrac{\beta}{h_{ie}'}$, $h_{ie}' = \dfrac{\beta}{g_m} = \dfrac{100}{0.1} = 1K$

g) $f_\beta = f_\tau/\beta = \dfrac{159 \text{Mhz}}{100} = 1.59 \text{Mhz}$

Solution 5:

a) $R_L = \dfrac{V_{CC} - V_{CE}}{I_C + I_B} = \dfrac{10 - 5}{2ma + 50\mu a} \approx 2.5K$

$R_B \approx \dfrac{V_{CE}}{I_b} = \dfrac{5V}{50\mu a} = 100K$

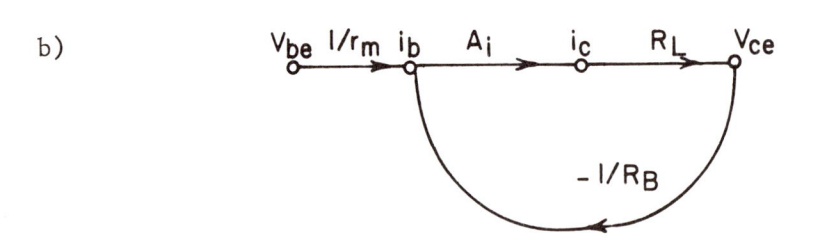

b)

c) $L = \dfrac{R_L}{r_{in}} A_i$, $L' = \dfrac{-R_L}{R_B} A_i$, $A'v = \dfrac{\dfrac{R_L}{r_{in}} A_i}{1 + \dfrac{R_L}{R_B} A_i}$, $A'i = \dfrac{A_i}{1 + \dfrac{R_L}{R_B} A_i}$

$db = 20 \log(1 + \dfrac{R_L}{R_B} A_i) = 20 \log(1 + \dfrac{2.5K}{100K} 40) = 6$

d) f'_2 is one octave above f_2

$f'_2 = (1 + \dfrac{R_L}{R_B} A_i) f_2 = 2 f_2$

FIRST SAMPLE P & P/EE EXAMINATION / 37

Solution 5 cont'd

e) $r'_{in} = \dfrac{r_{in}}{(1+\dfrac{R_L}{R_B}A_i)} = \dfrac{r_{in}}{2}$, $r_{in} = \dfrac{A_i R_L}{A_v} = \dfrac{40 \times 2.5K}{45} = 2.2K$

$r'_{in} = \dfrac{2.2K}{2} = 1.1K$

Solution 6:

$E = 100$ ft.c $A = (100)(200) = 2 \times 10^4$ ft^2

$N_L = 6$ lamps $\phi_L = 3,100$ lumens/lamp

C.U. = 0.78   M.F. = 0.6   M.H. = 19 ft.

(1) No. of 8' luminaires = $\dfrac{EA}{N_L \phi_L (C.U.)(M.F.)} = \dfrac{(10^2)(2 \times 10^4)}{6(3.1 \times 10^3)(0.78(0.6))} = 230$

(2) With long axis of machines in direction of plant length, orient rows in direction of plant width. The 100 ft. width allows rows of 11 eight foot luminaires and 1 four foot luminaire
(8 1/3' x 11+4 1/6' = 95.8')

Rows required = $\dfrac{230}{11.5} = 20$

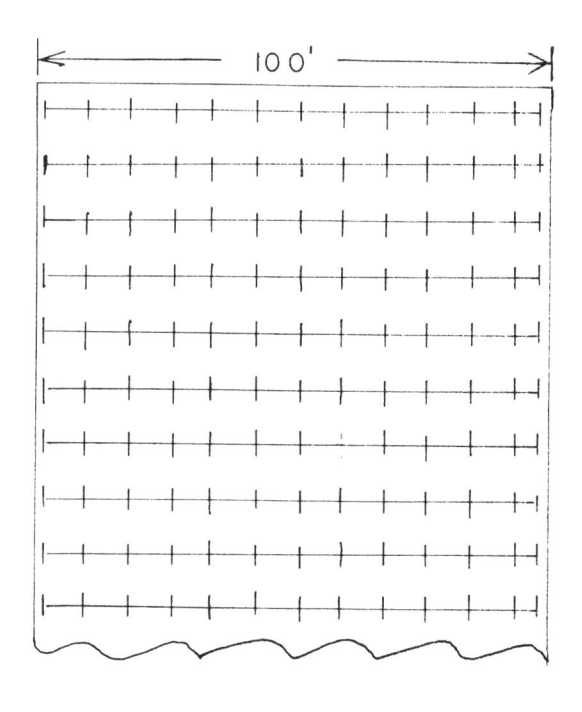

(3) 220 - 8' units
    20 - 4' units

Spacing:

Row center to center   10'
Row to end wall   5'
Row to side wall   2'1"

Hours/year = $17\dfrac{hrs}{day} \times 5\dfrac{day}{wk.}$

$\times 50\dfrac{wk.}{yr.} = 4,250$ hrs/yr.

Annual energy cost = 4250 (220)
(0.285) + (20)(0.190)(0.012)
= $3,391.50

Solution 7:

a) $R_T = \dfrac{\Delta°C}{\Delta mW} = \dfrac{(225-25)°C}{400mW} = \dfrac{200°C}{400mW} = 0.5°C/mW = R_T$

b) $R = \dfrac{E_{max}-V_D}{I_{max}} = \dfrac{10-1}{400ma}$ $(I_{max} = \dfrac{P_{max}}{V_{D_{max}}} = \dfrac{400mW}{1V} = 400ma)$

   $R = 22.5\Omega$

c) Derating Figure $= 1/R_T = \dfrac{1}{0.5°C/mW} = \dfrac{2mW}{°C}$

   $\Delta P = \Delta°C/R_T = (40-25)° \times \dfrac{2mW}{°C} = 30mW$

   P @ $40°C = (400-30)mW = 370mW$ $P = 370mW$

d) Derating $\Delta P = 400mW - 300mW = 100mW$

   $\Delta°C = \dfrac{P}{1/R_T} = \dfrac{100mW}{2.86mW/°C} = 35°C$ (above $25°C$)

   $T_A = (35+25) = 60°C$

Solution 8:

Standard form

$\ddot{e} + \dot{e} + e = 0$

$m^2 + m + 1 = 0$

$m = \dfrac{-1 \pm \sqrt{1^2 - 4}}{2}$ $\rightarrow$ Undamped

$e(t) = \varepsilon^{-\alpha t}(A\cos\omega_d t + B\sin\omega_d t)$

Sub values for α, ω_d

  From KCL $\Sigma i = 0$

$i_L + i_C + i_R = 0$

$\dfrac{1}{L}\int e\,dt + C\dfrac{de}{dt} + \dfrac{e}{R} = 0$

  Take Derivative

$e/L + C\ddot{e} + \dfrac{\dot{e}}{R} = 0$ $\ddot{e} + \dfrac{1}{RC}\dot{e} + \dfrac{1}{LC}e = 0$

Solution 8 cont'd

From Standard Form

$$\ddot{X} + 2\alpha \dot{X} + \omega_o^2 X = 0 \qquad 2\alpha = \frac{1}{RC} \qquad \alpha = \frac{1}{2RC} \qquad \alpha = \frac{1}{2\cdot 1 \cdot 1} = \frac{1}{2}$$

$$\alpha = \frac{1}{2}$$

$$\omega_o^2 = \frac{1}{LC} \qquad \omega_o = \frac{1}{\sqrt{LC}} = \frac{1}{1 \cdot 1} = 1$$

$$\omega_d = \sqrt{\omega_o^2 - \alpha^2} = \sqrt{1^2 - (\tfrac{1}{2})^2} = \sqrt{3/4}$$

$$\omega_d = \frac{\sqrt{3}}{2} = .866$$

$$e(t) = \varepsilon^{-t/2} (A \cos \tfrac{\sqrt{3}}{2} t + B \sin \tfrac{\sqrt{3}}{2} t)$$

To find A & B we need two I.C.'s $e(0)$ and $e'(0)$

1. $e(0) = \frac{1}{C} \int_{-\infty}^{t} i \, dt = \frac{1}{1} \int_{-0}^{+0} 6 U_0(t) \, dt = 6$

   $e(0) = 6 \qquad 6 = \varepsilon^{-0}(A \cos 0 + B \sin 0) \qquad 6 = A$

2. $e'(t) = \frac{de}{dt} = \frac{i_C}{C} = -\frac{e}{RC} \qquad e(0) = -\frac{6}{1 \cdot 1} = -6$

   $e'(t) = \varepsilon^{-t/2} (-\tfrac{\sqrt{3}}{2} A \sin \tfrac{\sqrt{3}}{2} t + \tfrac{\sqrt{3}}{2} B \cos \tfrac{\sqrt{3}}{2} t) + (-\tfrac{1}{2}) \varepsilon^{-t/2}$
   $(A \cos \tfrac{\sqrt{3}}{2} t + B \sin \tfrac{\sqrt{3}}{2} t)$

   $e'(0) = 1(0 + \tfrac{\sqrt{3}}{2} B) - \tfrac{1}{2} A + 0 = \tfrac{\sqrt{3}}{2} B - \tfrac{1}{2} \cdot 6 = \tfrac{\sqrt{3}}{2} B - 3$

   $-6 = \tfrac{3}{2} B - 3 \quad B = -3 \times \tfrac{2}{\sqrt{3}} = -3/\sqrt{6} \quad B = -3.46$

   $\therefore e(t) = \varepsilon^{-t/2} (6 \cos \tfrac{\sqrt{3}}{2} t - 3.46 \sin \tfrac{\sqrt{3}}{2} t)$

Solution 9: $I_L = (9+3+5) \text{mA} = 17 \text{ma}$

$I_{Bleed} = 0.1 \, I_L = 0.1 \times 17 \text{ma} = 1.7 \text{ ma}$

$I_T = I_L + I_B = (17 + 1.7) \text{ma} = 18.7 \text{ma}$

Solution 9 cont'd

$$R_3 = \frac{V_{R_3}}{I_T} = \frac{12V}{18.7\text{ma}} = 640$$

$$R_2 = \frac{V_{R_2}}{I_{BL}} = \frac{220}{1.7\text{ma}} = 130K$$

$$R_1 = \frac{V_{R_1}}{I_{BL}+I_{R_1}} = \frac{40}{6.7\text{ma}} = 6K$$

$$P_3 = V_{R_3} I_T = 12 \times 18.7\text{ma} = 225\text{mW}$$

$$P_2 = V_{R_2} I_B = 220 \times 1.7\text{ma} = 373\text{mW}$$

$$P_1 = V_{R_1}(I_B + I_{R_1}) = 40(5+1.7)\text{ma} = 268\text{mW}$$

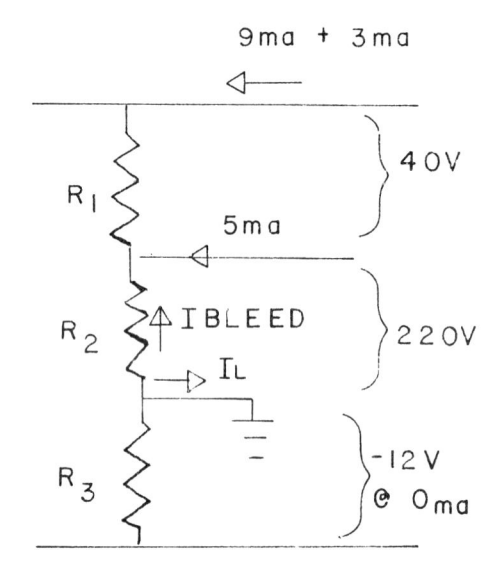

Part II

Illustrated Problems

Chapter 5

CIRCUITS

1) Design a low pass RC filter. This filter should have a half power frequency of 5,000 Hertz. The circuit configuration to be used is

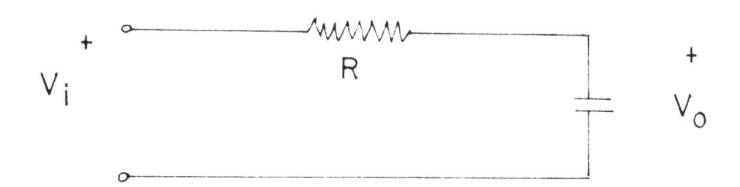

The transfer function for this filter is

$$\frac{V_o}{V_i} = \frac{1/RC}{S + 1/RC}$$

which, for sinusoidal excitation, becomes

$$\frac{V_o}{V_i} = \frac{1/RC}{j\omega + 1/RC}$$

The magnitude function $|V_o/V_i|$ has a half power frequency that occurs when $\omega = 1/RC$. Thus

$$1/RC = (5,000)(2)(\pi) = 3.14 \times 10^4$$

If we select $R = 1 \times 10^6 \, \Omega$, the capacitor should have the value of $31.8 \, \rho f = 31.8 \times 10^{-12} \, f$.

Now check!

$$\left|\frac{V_o}{V_i}\right|_{\omega = 3.14 \times 10^4} = \frac{\frac{1}{(1 \times 10^6)(31.8 \times 10^{-12})}}{j3.14 \times 10^4 + \frac{1}{(1 \times 10^6)(31.8 \times 10^{-12})}}$$

$$= \frac{1}{\sqrt{2}}$$

Which checks.

2. Another type of filter problem that can be encountered is that of designing an LC filter that is terminated in a resistor and driven by a current source. The configuration is shown below.

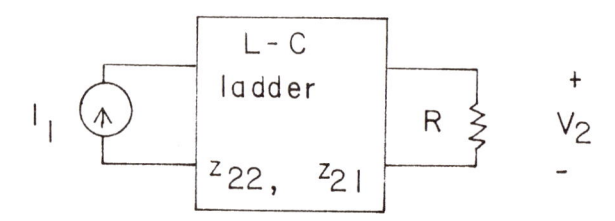

The transfer function of interest here is $Z_{21}(S)$ which can be written, for $R = 1$ ohm,

$$Z_{21} = \frac{z_{21}}{z_{22} + 1}$$

CIRCUITS / 45

Suppose we wish to design a 3rd Butterworth filter in the network structure shown below. The Butterworth polynomials can be found in any number of references.*

These show that the third order polynomial is

$$S^3 + 2S^2 + 2S + 1$$

and so we have

$$z_{21} = \frac{1}{S^3 + 2S^2 + 2S + 1} = \frac{z_{21}}{z_{22} + 1}$$

For this type of transfer function we divide the numerator and denominator by the odd part of the denominator polynomial. This leads to

$$z_{21} = \frac{\frac{1}{(S^3 + 2S)}}{\frac{2S^2 + 1}{S^3 + 2S} + 1} = \frac{z_{21}}{z_{22} + 1}$$

We can now identify

$$z_{22} = \frac{2S^2 + 1}{S^3 + 2S} \quad , \quad z_{21} = \frac{1}{S^3 + 2S}$$

The z_{21} has its transmission zeros at infinity thus we can realize the z_{22} as a ladder network of series inductors and shunt capacitors. A continued fraction expansion of z_{22} leads to

---

*Shea, <u>Amplifier Handbook</u>, pp. 5-9, McGraw-Hill Book Company. Weinberg, <u>Network Analysis and Synthesis</u>, p. 495, McGraw-Hill Book Company.

46 / CIRCUITS

$$
2S^2 + 1 \;\overline{\left)\begin{array}{l} \dfrac{S}{2} \leftarrow Y \\ S^3 + 2S \end{array}\right.}
$$

$$
S^3 + \dfrac{S}{2} \quad\overline{\left)\begin{array}{l} \dfrac{4}{3}S \leftarrow Z \\ \dfrac{3}{2}S \;\overline{\big)\; 2S^2 + 1} \end{array}\right.}
$$

$$
2S^2 \quad\overline{\left)\begin{array}{l} \dfrac{3}{2}S \leftarrow Y \\ 1 \;\overline{\big)\; \dfrac{3}{2}S} \\ \dfrac{3}{2}S \end{array}\right.}
$$

and the network becomes

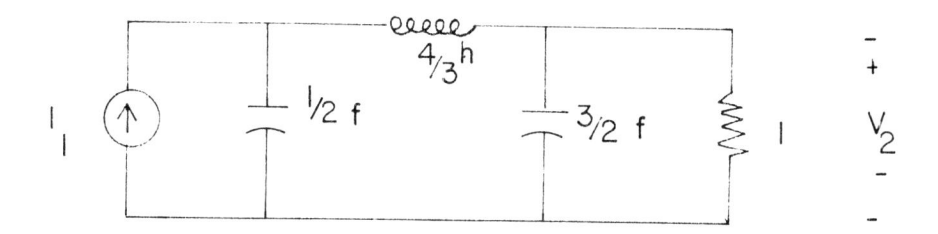

[Circuit diagram: current source I_1 in parallel with $\tfrac{1}{2}$ f capacitor, then $\tfrac{4}{3}$ h inductor in series, then $\tfrac{3}{2}$ f capacitor in parallel with 1 ohm resistor across which V_2 is measured.]

If the terminating resistor is to be other than one ohm, **resistance** scaling may be utilized. Suppose, for example, we wish to have the terminating resistor equal to 600 ohms. To accomplish this, we multiply all resistors and industors by 600 and divide the capacitor by 600. The above filter has a half power frequency of $\omega = 1$. We can "frequency scale" the network so as to have the half power frequency located at any desired value. Suppose the specified half power frequency is to be $\omega = 10{,}000$ r/s. Then we would divide all inductor and capacitor values by 10,000.

If we combine these two parameter adjustment methods, we have the new values of

$$R = (1)(600) = 600 \, \Omega$$

$$L = \frac{(1.333)(600)}{10,000} = 79.98 \text{ mh}$$

$$C_1 = \frac{0.5}{(600)(10,000)} = 0.083 \, \mu f$$

$$C_2 = \frac{1.5}{(600)(10,000)} = 0.25 \, \mu f$$

The final filter is shown below.

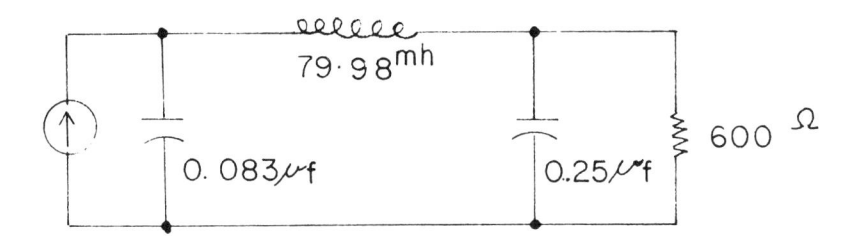

3) As an illustration of a transient problem consider the circuit shown below. The switch has been in position 1 for a very long time and then is thrown to position 2. Find the current in the inductor after the switch is thrown.

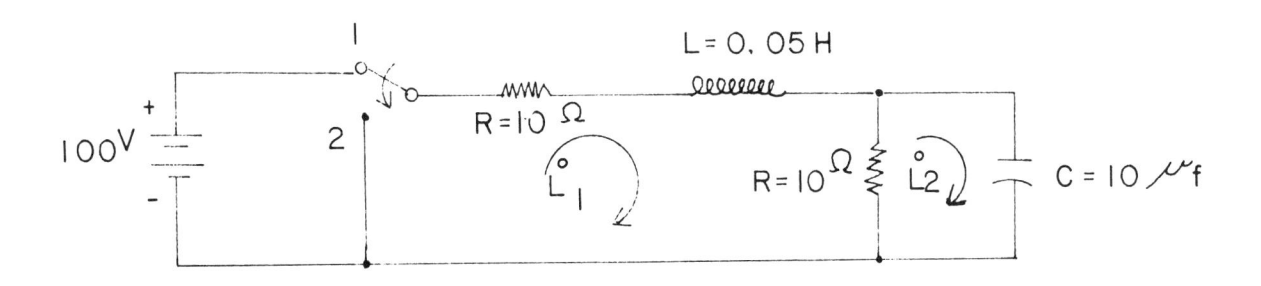

48 / CIRCUITS

First find the initial conditions. The assumption made here is that the switch has been in position 1 long enough so that a steady state condition has been reached. This being the case, the initial current through the inductor is given by

$$I_L(0^+) = \frac{100}{10 + 10} = 5 \text{ amp}$$

and the voltage across the capacitor will be equal to the voltage across the right-hand resistor which is equal to 50 volts.

This problem will be solved with the aid of the Laplace transform. First, write the two loop equations for the time interval after the switch is moved to position 2. These equations are

$$Ri_1 + L\frac{di}{dt} + R(i_1 - i_2) = 0$$

and

$$\frac{1}{C}\int i_2 \, dt + R(i_2 - i_1) = 0$$

Transforming these two equations yields

$$RI_1(S) + SLI_1(S) - i_1(0) + R(I_1(S) - I_2(S)) = 0$$

and

$$\frac{1}{C} I_2(S) + \frac{V_C(0)}{S} + R(I_2(S) - I_1(S)) = 0$$

Substituting the values of the initial conditions and separating variables leads to

$$0.25 = I_1(R + R + SL) - I_2(R)$$

$$\frac{-50}{S} = -I_1(R) + I_2(R + 1/CS)$$

These two equations may now be solved for I_1.

If Cramer's Rule is used we have

$$I_1 = \frac{\begin{vmatrix} 0.25 & -R \\ -\frac{50}{S} & R + \frac{1}{CS} \end{vmatrix}}{\begin{vmatrix} 2R + SL & -R \\ -R & R + \frac{1}{CS} \end{vmatrix}}$$

$$= \frac{0.25(R + \frac{1}{CS}) - \frac{50R}{S}}{2R^2 + \frac{2R}{CS} + RSL + \frac{L}{C} - R^2}$$

$$= \frac{0.25R + \frac{0.25}{CS} - \frac{50R}{S}}{R^2 + \frac{2R}{CS} + RSL + \frac{L}{C}}$$

$$= \frac{0.25RS + \frac{0.25}{C} - 50R}{S^2RL + S\{\frac{L}{C} + R^2\} + \frac{2R}{C}}$$

$$= \frac{0.25R(S + \frac{1}{RC} - \frac{50}{0.25})}{RL(S^2 + S[\frac{1}{RC} + \frac{R}{L}] + \frac{2}{LC})}$$

Substituting the numerical values leads to

$$I_1 = 5\left\{\frac{S + 9.8 \times 10^3}{S^2 + 1.02 \times 10^4 S + 4 \times 10^6}\right\}$$

$$I_1 = 5\left\{\frac{S + 9.8 \times 10^3}{(S + 4.085 \times 10^2)(S + 9.791 \times 10^3)}\right\}$$

50 / CIRCUITS

The value of $i_1(t)$, the inductor current, can be obtained by taking the inverse Laplace transform of the above equation. To accomplish this, first perform a partial fraction expansion of the expression for I_1. That is, we wish to write I_1 in the form

$$I_1 = \frac{K_1}{S + 4.085 \times 10^2} + \frac{K_2}{S + 9.791 \times 10^3}$$

Now,

$$K_1 = (S + 4.085 \times 10^2) I_1 \bigg|_{S = -4.085 \times 10^2}$$

$$= \frac{5(S + 9.8 \times 10^3)}{S + 9.791 \times 10^3} \bigg|_{S = -4.085 \times 10^2}$$

$$= \frac{5(-4.085 \times 10^2 + 9.8 \times 10^3)}{-4.085 \times 10^2 + 9.791 \times 10^3}$$

$$= \frac{5(9.392 \times 10^3)}{9.383 \times 10^3}$$

$$= 5.005$$

and

$$K_2 = (S + 9.791 \times 10^3)I_1 \Big|_{S = -9.791 \times 10^3}$$

$$= -4.796 \times 10^{-3}$$

Thus

$$I_1 = \frac{5.005}{S + 4.085 \times 10^2} - \frac{4.796 \times 10^{-3}}{S + 9.791 \times 10^3}$$

and

$$i_1(t) = 5.005\, e^{-(408.5)t} - 0.004796\, e^{-(9791)t}$$

4) For the circuit below, find $V_o(j\omega)$ and comment upon the behavior of the network as a filter.

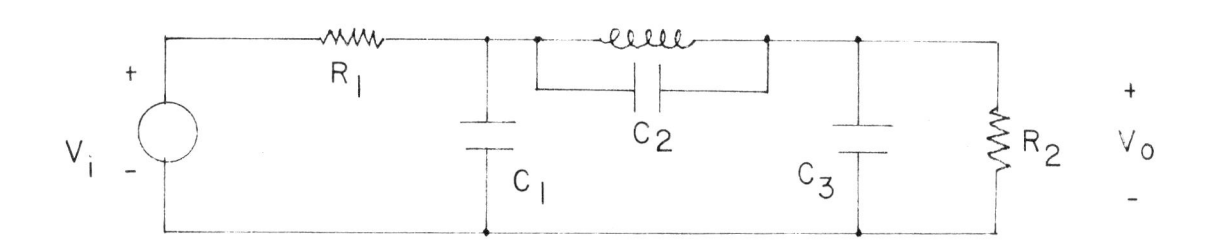

52 / CIRCUITS

This problem will be approached by performing a source transformation and then writing nodal equations. All variables and impedances will be written in Laplace transform notation (to reduce the complex algebra manipulation). When the solution is finally obtained, S will be replaced by $j\omega$ thus yielding the desired result.

The transformed network is shown below.

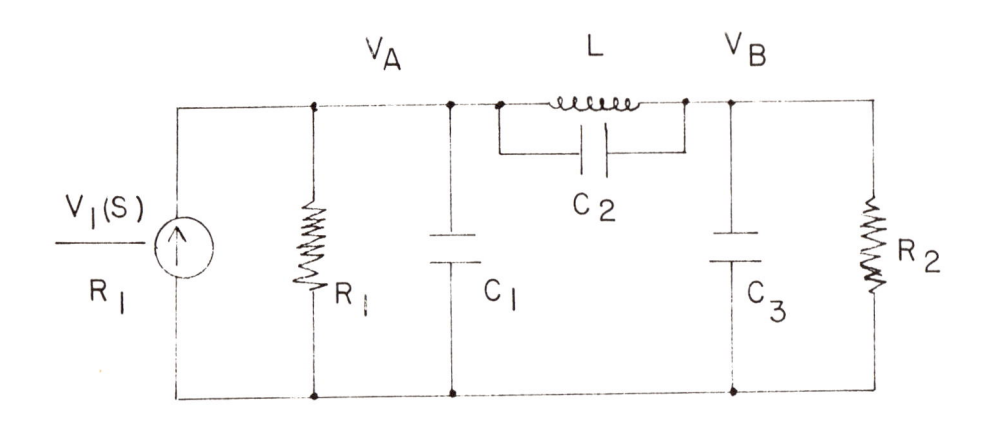

The nodal equations are

$$\frac{V_i(S)}{R_1} = V_A(G_1 + C_1 S + C_2 S + \frac{1}{SL}) - V_B(C_2 S + \frac{1}{SL})$$

$$0 = -V_A(C_2 S + \frac{1}{SL}) + V_B(G_2 + C_2 S + C_3 S + \frac{1}{SL})$$

Now,

$$V_B = V_o = \frac{\begin{vmatrix} G_1 + C_1 S + C_L S + \frac{1}{SL} & \frac{V_i(S)}{R_1} \\ -(C_2 S + \frac{1}{SL}) & 0 \end{vmatrix}}{\begin{vmatrix} G_1 + C_1 S + C_2 S + \frac{1}{SL} & -(C_2 S + \frac{1}{SL}) \\ -(C_2 S + \frac{1}{SL}) & G_2 + C_2 S + C_3 S + \frac{1}{SL} \end{vmatrix}}$$

$$= \frac{\left(\frac{V_i(S)}{R_1}\right)(C_2 S + \frac{1}{SL})}{S^2(C_1C_2 + C_1C_3 + C_2C_3) + S(G_2C_1 + G_2C_2 + G_1C_2 + G_1C_3) + G_1G_2 + \frac{1}{L}(C_1 + C_3) + \frac{1}{SL}(G_1 + G_2)}$$

$$= \frac{1}{R_1 L}\left\{\frac{V_i(S)[S^2 C_2 L + 1]}{S^3(C_1C_2 + C_1C_3 + C_2C_3) + S^2(G_1[C_2 + C_3] + G_2[C_1 + C_2]) + S(G_1G_2 + \frac{C_1 + C_3}{L}) + \frac{G_1 + G_2}{L}}\right\}$$

The next step is to convert this function from $V_o(S)$ to $V_o(j\omega)$. Doing this leads to

$$V_o(j\omega) = \frac{V_i(j\omega)[1 - \omega^2 C_2 L]/R_1 L}{\frac{G_1 + G_2}{L} - \omega^2\{G_1[C_2 + C_3] + G_2[C_1 + C_2]\} + j\{\omega(G_1 G_L + \frac{C_1 + C_3}{L}) - \omega^3(C_1C_2 + C_1C_3 + C_2C_3)\}}$$

With regard to the filtering action provided by the network, it can be seen that for $V_o(j\omega)/V_i(j\omega)$

    a) $V_o(j\omega) \to 0$ as $\omega \to \infty$

    b) $V_o(j\omega) \to K$ as $\omega \to 0$

c) The numerator becomes zero when

$$1 - \omega^2 C_2 L = 0 \qquad \text{or} \qquad \omega = 1/\sqrt{C_2 L}$$

Thus the network behaves as a low pass filter with a notch type of response at $\omega = 1/\sqrt{C_2 L}$. In other words, the network has a "transmission zero" at $\omega = 1/\sqrt{C_2 L}$.

5) The next few questions have to do with resonant circuits and bandwidth. First consider the parallel circuit below. It is called a parallel resonant network.

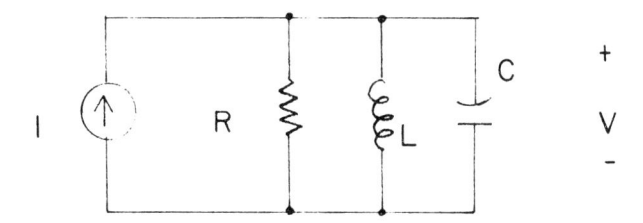

Resonance occurs (the network is said to be in resonance) when the voltage and current at the input terminals are in phase. The admittance offered to the current source is

$$Y = \frac{1}{R} + j(\omega C - \frac{1}{\omega L})$$

and thus resonance occurs when

$$\omega C - \frac{1}{\omega L}$$

and the resonant frequency is

$$\omega_o = 1/\sqrt{LC} \qquad \text{or} \qquad f_o = 1/2\pi\sqrt{LC}$$

Two other parameters that are often encountered in the discussion of resonant circuits are the equality factor, Q, and the bandwidth, B. The Q is defined as

$$Q = 2\pi \frac{\text{maximum energy stored}}{\text{total energy lost per period}}$$

At resonance Q can be shown to be equal to

$$Q = \omega_o RC = R/\omega_o L$$

The bandwidth is defined as the difference between the two half power frequencies ω_2 and ω_1. That is,

$$B = \omega_2 - \omega_1$$

The half power frequencies are those frequencies at which the input admittance of the circuit has a magnitude that is larger than that at resonance by a factor of $\sqrt{2}$.

Some relations between these two parameters are

$$\omega_2 = \omega_o \left[\sqrt{1 + \left(\frac{1}{2Q}\right)^2} + \frac{1}{2Q} \right]$$

$$\omega_1 = \omega_o \left[\sqrt{1 + \left(\frac{1}{2Q}\right)^2} - \frac{1}{2Q} \right]$$

and

$$B = \frac{\omega_o}{2Q}$$

If the circuit has a Q of 5 or more, then the following approximation holds:

$$\omega_{1,2} \approx \omega_o \mp \frac{B}{2}$$

56 / CIRCUITS

As an example, suppose we have a parallel resonant circuit with R = 30 KΩ, L = 50 mH, and C = 0.25 μf. The circuit has the following parameter values:

$$\omega_o = 1/\sqrt{LC} = 8.944 \times 10^3 \text{ r/s}$$

$$f_o = \frac{\omega_o}{2\pi} = 1.424 \times 10^3 \text{ hertz}$$

$$Q = \omega_o RC = (8.944 \times 10^3)(30 \times 10^3)(.25 \times 10^{-6}) = 67.1$$

The value of Q, in this case, is large enough so that the approximation may be used to find the upper and lower half power frequencies. Thus,

$$B = \frac{\omega_o}{2Q} = \frac{8.944 \times 10^3}{(2)(67.1)} = 6.66 \times 10^1 \text{ r/c}$$

and

$$\omega_1 = 8944 - 66.6/2 = \mathbf{8,911 \text{ r/c}}$$

$$\omega_2 = 8944 + 66.6/2 = \mathbf{8,977 \text{ r/c}}$$

6) It is also possible to encounter series resonant circuits. One is shown below.

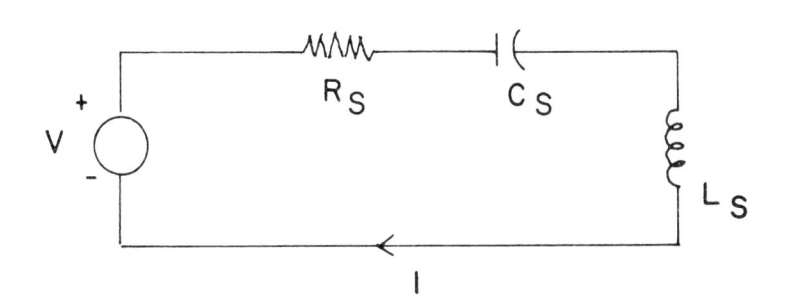

The impedance of this circuit is

$$Z = R_s + j(\omega L_s - \frac{1}{\omega C_s})$$

and again resonance occurs when

$$\omega L_s - \frac{1}{\omega C_s} = 0$$

or

$$\omega_{os} = \frac{1}{\sqrt{L_s C_s}}$$

Proceeding as in the case of the parallel resonant circuit we have

$$Q_{os} = \frac{\omega_{os} L_s}{R_s}$$

$$B_s = \omega_{2s} - \omega_{1s} = \frac{\omega_{os}}{Q_{os}}$$

and for high Q circuits

$$\omega_{1s,2s} \approx \omega_{os} \mp \frac{B_s}{2}$$

As an example, suppose we wish to find the bandwidth of a series RLC circuit having component values R = 10 Ω, L = 50 mH, C = 5 μf.

$$\omega_{os} = 1/\sqrt{L_s C_s} = 2,000 \text{ r/s}$$

$$f_{os} = \frac{\omega_{os}}{2\pi} = 318.3 \text{ hertz}$$

$$Q_{os} = \frac{\omega_{os} L_s}{R_s} = 10$$

$$B = \frac{\omega_{os}}{Q_{os}} = \frac{2,000}{10} = 200 \text{ r/s}$$

58 / CIRCUITS

$$\omega_{1s} = \omega_{os} - \frac{B_s}{2} = 2{,}000 - \frac{200}{2} = 1{,}900 \text{ r/s}$$

$$\omega_{2s} = \omega_{os} + \frac{B_s}{2} = 2{,}000 + \frac{200}{2} = 2{,}100 \text{ r/s}$$

7) Frequently the circuits encountered in practice are neither the simple series or parallel type that were just discussed. A more practical example is shown below.

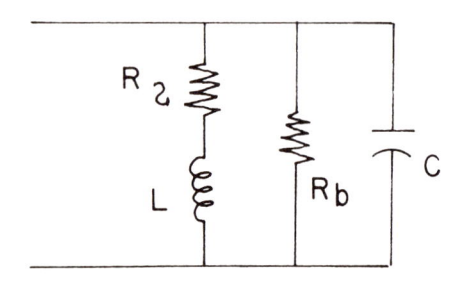

It is generally possible to deal with a simpler circuit than that above, this simpler circuit being equivalent to the given circuit over a band of frequencies that is usually large enough to include all frequencies of interest.

To see how this is done, consider the two circuits shown below.

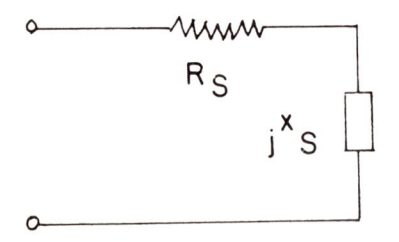

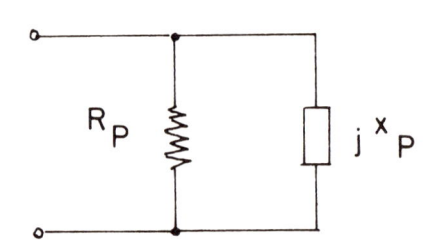

For these circuits we may define Q as was done previously and this may be done at any frequency, although we will only deal with the resonant frequency of the network of which these arms are parts. The Q of the series arm is $|X_s|/R_s$ while the Q of the parallel network is $R_p/|X_p|$. In order for the two networks to be equivalent, it is necessary that the following relations be satisfied.

$$R_p = R_s(1 + Q^2)$$

$$X_p = X_s(1 + \frac{1}{Q^2})$$

where $Q_p = Q_s = Q$.

Thus if we know R_s and X_s we may replace the series circuit with and equivalent parallel circuit. If Q is greater than 5, the following approximations may be used.

$$R_p \approx Q^2 R_s$$

$$X_p \approx X_s \qquad (C_p \approx C_s \text{ or } X_p \approx X_s)$$

As an example of finding an equivalent circuit in this manner, consider the circuit below and find the parallel equivalent to it.

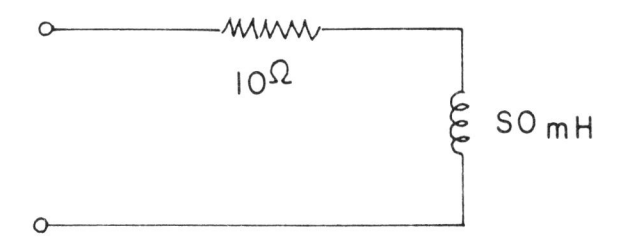

We wish to find the equivalent at a specific frequency, e.g., $\omega = 2,000$ r/s. At this frequency

$$Q_s = \frac{\omega L_s}{R_s} = \frac{(2,000)(50 \times 10^{-3})}{10} = 10$$

and

$$R_p = R_s(1 + Q^2) = 10(1 + 10^2) = 10(101) = 1,010$$

$$L_p = L_s(1 + \frac{1}{Q^2}) = 50 \times 10^{-3}(1 + \frac{1}{10^2}) = 5.05 \times 10^{-2}$$

$$= 50.5 \text{ mH}$$

Let us check the Q of the parallel and the impedance at $\omega = 2,000$ r/s of the two networks.

$$Q_p = \frac{R_p}{\omega L_p} = \frac{1,010}{(2,000)(50.5 \times 10^{-3})} = 10$$

$$Z_s = 10 + j(2,000)(50 \times 10^{-3}) = 10 + j100 = 100.5 \underline{/84.29°}$$

$$Z_p = \frac{1}{Y} = \frac{1}{\frac{1}{1,010} + \frac{1}{j(2,000)(50.5 \times 10^{-3})}}$$

$$= \frac{1}{9.901 \times 10^{-4} - j9.901 \times 10^{-3}}$$

$$= \frac{1}{9.95 \times 10^{-3} \underline{/-84.29°}} = 100.5 \underline{/84.29°}$$

Thus it is seen that the two networks are indeed equivalent at the frequency $\omega = 2,000$ r/s.

8) As another example of the use of this equivalent circuit concept, suppose we wish to analyze the network shown below.

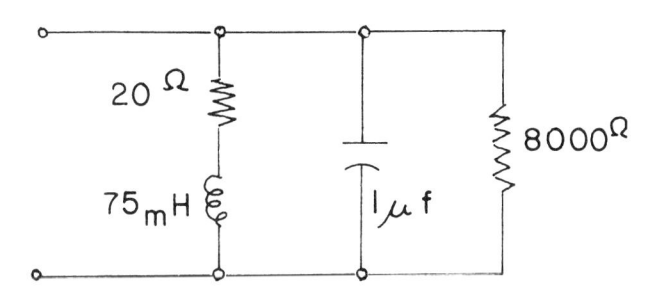

This problem will be solved by replacing the series resistor-inductor branch by its parallel equivalent. This will be done at the resonant frequency given by

$$\omega_o = \frac{1}{\sqrt{(1 \times 10^{-6})(75 \times 10^{-3})}} = 3.651 \times 10^3 \text{ r/s}$$

This frequency is not exactly the resonant frequency of the actual network but in almost all cases of interest the Q of the circuit is high enough to allow this approximation to be used. At this frequency, the series arm, Q_s, is

$$Q_s = \frac{(3.651 \times 10^3)(75 \times 10^{-3})}{20} = 13.69$$

Now,

$$R_p = 20(1 + 13.69) = 3.77 \text{ K}$$

$$L_p = 75 \times 10^{-3}(1 + \frac{1}{13.69^2}) = 75.4 \text{ mH}$$

62 / CIRCUITS

Thus we may consider the network

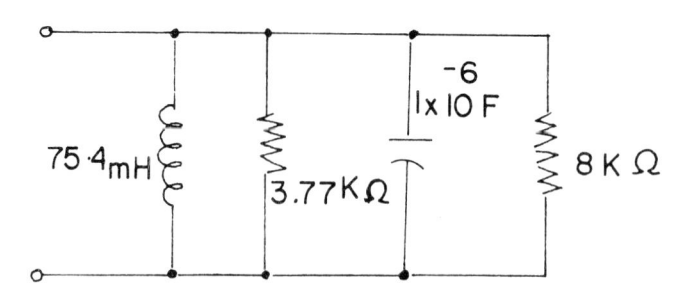

[Circuit diagram: 75.4 mH inductor, 3.77 KΩ resistor, 1×10⁻⁶ F capacitor, and 8 KΩ resistor in parallel]

and the two resistors may be combined. If this is done, the resulting circuit is

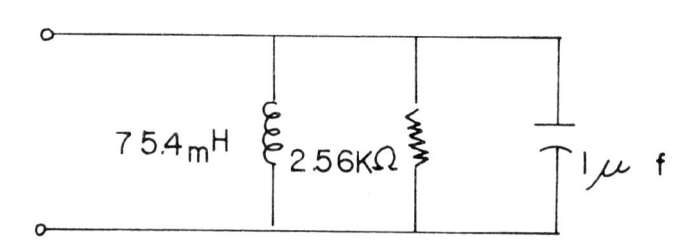

[Circuit diagram: 75.4 mH inductor, 2.56 KΩ resistor, and 1 μf capacitor in parallel]

This network may now be analyzed and the results will be a close approximation to the corresponding properties of the initial circuit. Therefore,

$$\omega_o = \frac{1}{\sqrt{(75.4 \times 10^{-3})(1 \times 10^{-6})}} = 3.642 \times 10^3 \text{ r/s}$$

$$Q = (3.642 \times 10^3)(2.56 \times 10^3)(1 \times 10^{-6}) = 9.3$$

$$B = \frac{3.642 \times 10^3}{(2)(9.3)} = 195.8 \text{ r/s}$$

and

$$\omega_{1,2} \approx 3.642 \times 10^3 \mp 195.8/2 \text{ r/s}$$

9) The need to deal with networks containing mutual inductances occurs quite frequently. As an example, consider the network below.

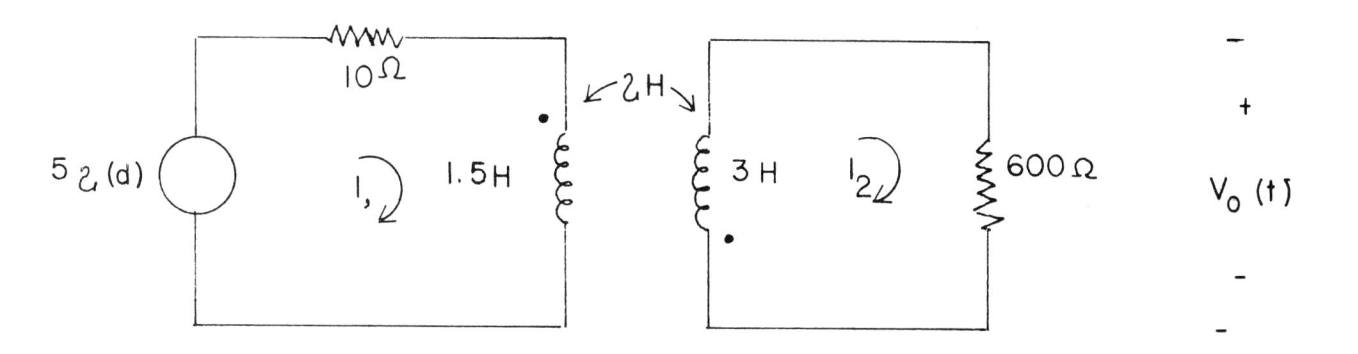

We wish to find the voltage across the 600 ohm resistor when a 5 volt step function is applied to the input. Using Laplace transform methods and assuming there are no initial conditions, we have

$$\frac{5}{s} = I_1(s)\{10 + 1.5s\} + I_2(s)\{2s\}$$

$$0 = I_1(s)\{2s\} + I_2(s)\{3s + 600\}$$

or, using matrix notation we have

$$\begin{bmatrix} \frac{5}{s} \\ 0 \end{bmatrix} = \begin{bmatrix} 10 + 1.5s & 2s \\ 2s & 3s + 600 \end{bmatrix} \begin{bmatrix} I_1(s) \\ I_2(s) \end{bmatrix}$$

To find the voltage across the 600 Ω resistor we need to determine the current $I_2(s)$. Therefore, using matrix notation,

$$\begin{bmatrix} I_1(s) \\ I_2(s) \end{bmatrix} = \begin{bmatrix} 10 + 1.5s & 2s \\ 2s & 3s + 600 \end{bmatrix}^{-1} \begin{bmatrix} \frac{5}{s} \\ 0 \end{bmatrix}$$

Now

$$\begin{bmatrix} 10 + 1.5s & 2s \\ 2s & 3s + 600 \end{bmatrix}^{-1} = \frac{\begin{bmatrix} 3s + 600 & -2s \\ -2s & 10 + 1.5s \end{bmatrix}}{(10 \times 1.5s)(3s + 600) - (2s)^2}$$

$$= \frac{\begin{bmatrix} 3s + 600 & -2s \\ -2s & 10 + 1.5s \end{bmatrix}}{30s + 6{,}000 + 4.5s^2 + 900s - 4s^2}$$

$$= \frac{\begin{bmatrix} 3s + 600 & -2s \\ -2s & 10 + 1.5s \end{bmatrix}}{0.5s^2 + 930s + 600}$$

With this result we can find the two currents by performing the following matrix multiplication.

$$\begin{bmatrix} I_1(s) \\ I_2(s) \end{bmatrix} = \frac{\begin{bmatrix} 3s + 600 & -2s \\ -2s & 1.5s + 10 \end{bmatrix} \begin{bmatrix} \frac{5}{s} \\ 0 \end{bmatrix}}{0.5s^2 + 930s + 600}$$

$$= \frac{\begin{bmatrix} \frac{5}{s}(3s + 600) + (-2s)(0) \\ \frac{5}{s}(-2s) + (1.5s + 10)(0) \end{bmatrix}}{0.5s^2 + 930s + 600}$$

and

$$I_2(s) = \frac{5}{s}(-2s)/(0.5s^2 + 930s + 600)$$

$$= \frac{-10}{0.5s^2 + 930s + 600}$$

$$= \frac{-20}{s^2 + 1,860s + 1,200} = \frac{-20}{(s + .645)(s + 1,859.4)}$$

$$= \frac{0.01076}{s + 1,859.4} - \frac{0.01076}{s + .645}$$

Taking the inverse Laplace transform yields

$$i_2(t) = 0.01076\, e^{(-1,859.4)t} - 0.01076\, e^{(-0.645)t}$$

and finally the voltage across the 600 ohm resistor is given by

$$v_o(t) = 600\, i_2(t)$$

$$= 6.45\, e^{(-1,859.4)t} - 6.45\, e^{(-0.645)t}$$

(10) It is possible to replace a mutual inductance with a T equivalent circuit. That is, given the circuit below

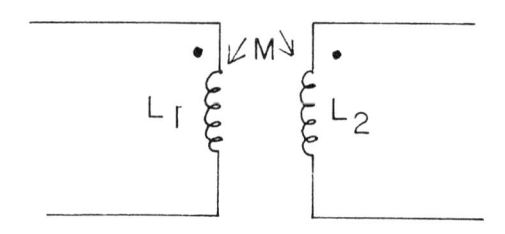

we may replace it with the following network.

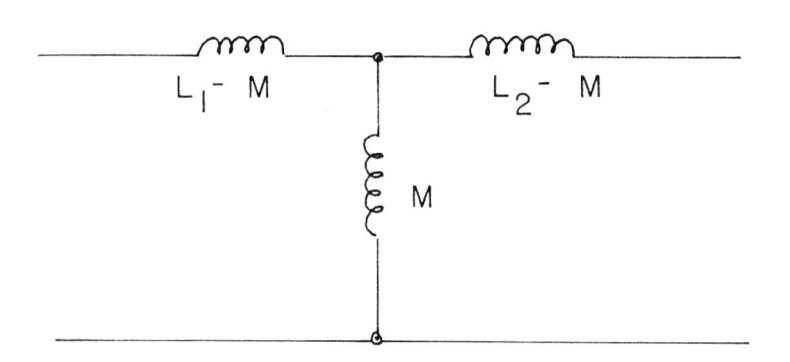

Note that if either of the dots on the windings of the mutual inductance are located on the opposite ends of the coils then the M in the equivalent circuit is replaced by −M.

As an example, consider the network analyzed in the previous problem. If the mutual inductance in that circuit is replaced by its T equivalent we have

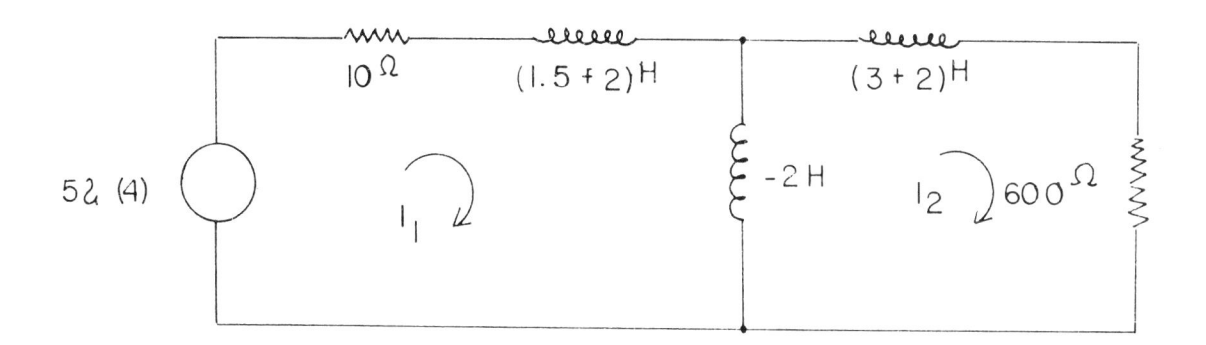

If the loop equations for this circuit are now written, we have

$$\frac{5}{s} = I_1(s)\{10 + 3.5s - 2s\} + I_2\{2s\}$$

$$0 = I_1(s)\{2s\} + I_2(s)\{600 + 5s - 2s\}$$

and these, of course, are the same equations we obtained earlier.

Chapter 6

CONTROLS

Problem 1

This is an angle transmitting system. The transmitter itself is the forward block having a system function of $K/s(1 + \tau s)$. It is incorporated into a unity feedback system to be analyzed. The diagram is shown below.

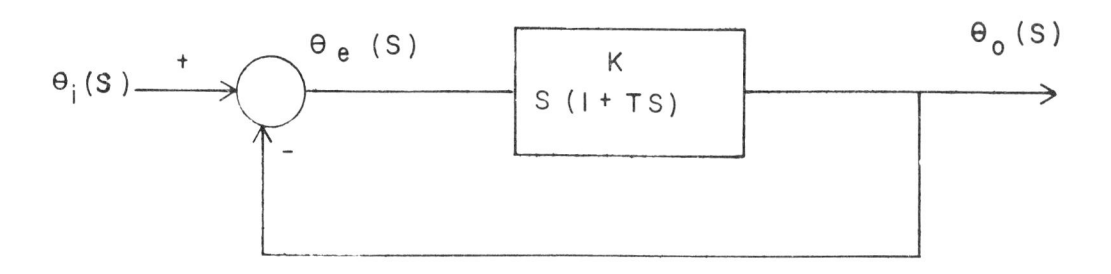

```
                          θ_e(S)      ┌──────────┐
θ_i(S) ────+──→( )──────────────────→ │    K     │ ──────→ θ_o(S)
            -  ↑                      │ S(1+TS)  │
               │                      └──────────┘
               │                            │
               └────────────────────────────┘
```

where $\theta_i(s)$ is the input angle

$\theta_o(s)$ is the output angle

$\theta_e(s)$ is the error angle

70 / CONTROLS

<u>Question 1.1</u>: Find the steady state error for a step input.

Write down what the error angle is:

$$\theta_e(s) = \theta_i(s) - \theta_o(s)$$

The system transfer function, $H(s)$, is:

$$H(s) = \frac{\theta_o(s)}{\theta_i(s)} = \frac{\text{forward function}}{1 + \text{loop function}}$$

where forward function $= G(s) = \dfrac{K}{s(1 + \tau s)}$

Therefore, the error angle is:

$$\theta_e(s) = \theta_i(s)[1 - H(s)]$$

or $\theta_e(s) = \theta_i(s)[1 - \dfrac{G(s)}{1 + G(s)}]$

or $\theta_e(s) = \theta_i(s)[\dfrac{1}{1 + G(s)}]$

The steady state error, θ_{ess}, is:

$$\theta_{ess} = \lim_{t \to \infty} \theta_e(t) = \lim_{s \to 0} s\theta_e(s)$$

The step input has a transform of $1/s$.

Therefore, after making substitutions:

$$\theta_{ess} = \lim_{s \to 0} s \, \frac{1/s}{1 + K/s(1 + \tau s)} = 0$$

Question 1.2: Find the steady state error for a ramp input.

A ramp input has a transform of $1/s^2$.

Therefore, as before

$$\theta_{ess} = \lim_{s \to 0} s \frac{1/s^2}{1 + K/s(1 + \tau s)} = 1/K$$

Question 1.3: For this angle transmitting system, find the value of forward gain, K, so that the steady state error, θ_{ess}, is less than or equal to one milliradian for a ramp input.

$$\theta_{ess} = 1/K \leq 10^{-3} \text{ radian}$$

or $K \geq 1,000$

Question 1.4: Given the following system specifications:

- the percent overshoot, i.e., P.O. $\leq 5\%$
- the settling time, T_s, < 4 sec.
- the rise time, T_r, < 1 sec.

for a step input.

Find the allowable area in the s-plane for the system poles.

It can be verified from any control theory text that a second order system has the following system function:

$$H(s) = \frac{\omega_n^2}{s^2 + 2\xi\omega_n s + \omega_n^2}$$

and for a P.O. = 5%, $\xi = .707$

and the settling time, $T_s = 4/\xi\omega_n$

and the rise time, $T_r = \pi/\omega_n \sqrt{1 - \xi^2}$

The characteristic equation holds the information on the position of the poles. It is found in the denominator of the system function. It is:

$$s^2 + 2\xi\omega_n s + \omega_n^2$$

The roots of this equation are

$$s = -\xi\omega_n \pm j\omega_n\sqrt{1 - \xi^2}$$

The values that s is allowed to take are the positions that the poles are allowed in the s-plane. An s-plane is shown below:

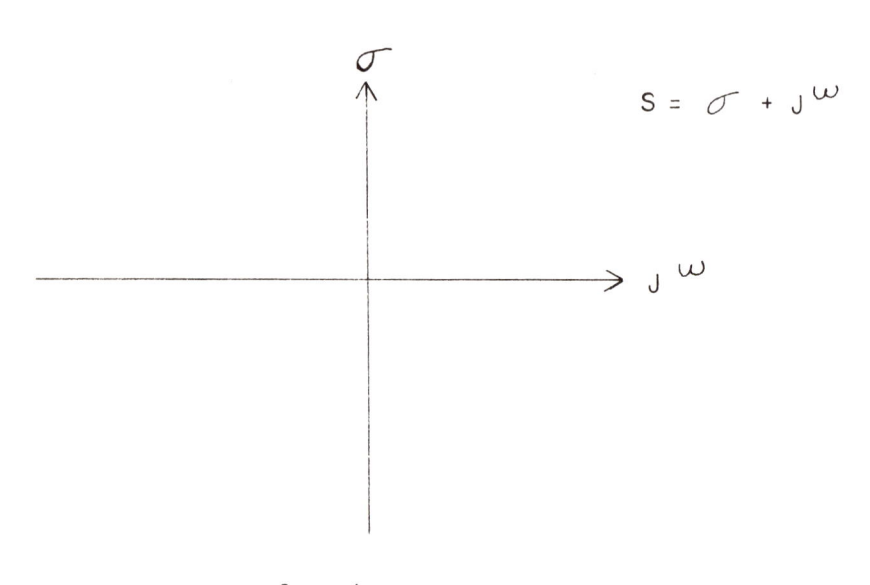

$$S = \sigma + j\omega$$

S - plane

The specifications are:

$$T_s = 4/\xi\omega_n < 4$$

$$T_r = \pi/\omega_n\sqrt{1-\xi^2} < 1$$

$$P.O. \leq 5\% \text{ or } \xi \geq .707$$

Therefore:

$$\xi\omega_n > 1$$

$$\omega_n\sqrt{1-\xi^2} > \pi$$

$$\xi \geq .707$$

Now the real part of the system poles, σ, is:

$$\sigma = -\xi\omega_n > -1$$

And the imaginary part of the system poles, ω, is:

$$\pm\omega = \pm\omega\sqrt{1-\xi^2} > \pm\pi$$

Finally, the angle of the complex pole, θ, is:

$$\theta = \tan^{-1}\omega/\sigma = \tan^{-1}\sqrt{1-\xi^2}/\xi$$

For $\xi = .707$, $\theta = \tan^{-1} 1 = 45^0$

These three pieces of information define the allowable area in the s-plane for the system poles as shown below:

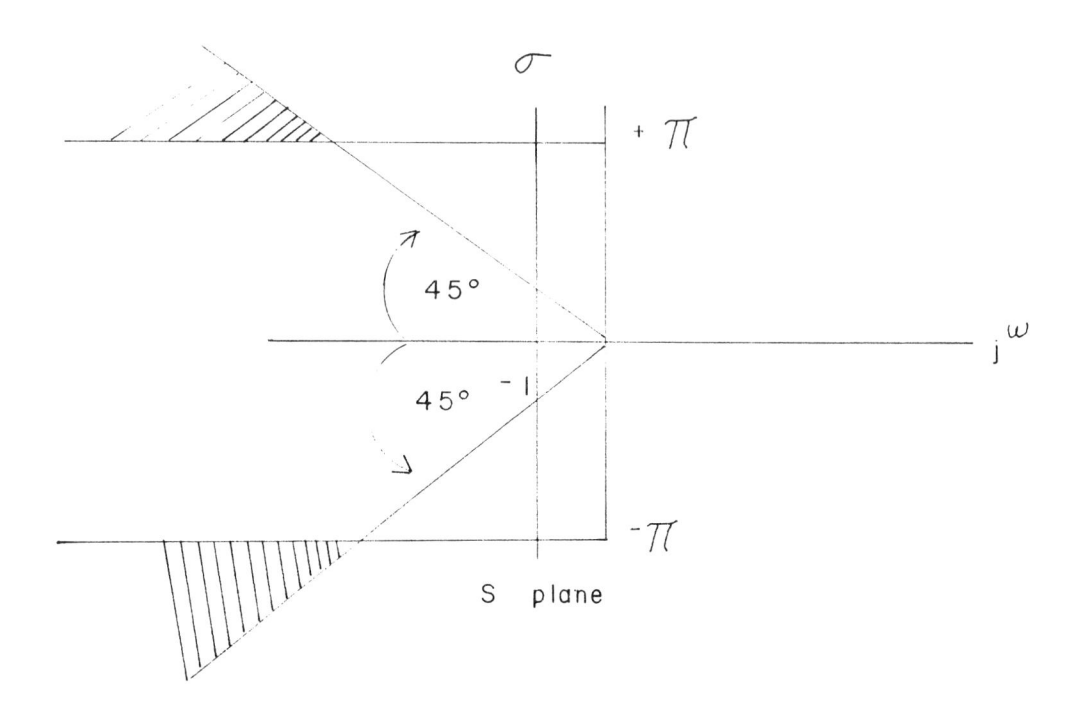

S plane diagram with ±45° lines from −1 and vertical bounds at ±π

Question 1.5: For a gain K = 1,000 and damping coefficient $\xi = .707$, find the value of τ which places the poles within the allowable area.

Write down the system function, $\theta_o(s)/\theta_i(s)$.

$$\theta_o(s)/\theta_i(s) = \frac{K/s(1+\tau s)}{1 + K/s(1+\tau s)} = \frac{K/\tau}{s^2 + \frac{1}{\tau}s + K/\tau}$$

The standard second order system has the form:

$$H(s) = \frac{\omega_n^2}{s^2 + 2\xi\omega_n s + \omega_n^2}$$

By equating similar coefficients of s:

$$\omega_n^2 = K/\tau$$

$$2\xi\omega_n = 1/\tau$$

Now substitute the expression for ω_n from the second equation into the first.

$$\frac{1}{4\xi^2\tau^2} = K/\tau \quad \text{or} \quad \tau = \frac{1}{4\xi^2 K}$$

Using the numerical values leads to

$$\tau = .5 \text{ ms}$$

Question 1.6: Locate the poles of question 1.5 to show that they are in the prescribed area in the s-plane.

The real part is $\sigma = -\xi\omega_n = 1,000$

The imaginary part is $\omega = \omega_n\sqrt{1 - \xi^2} = 1,000$

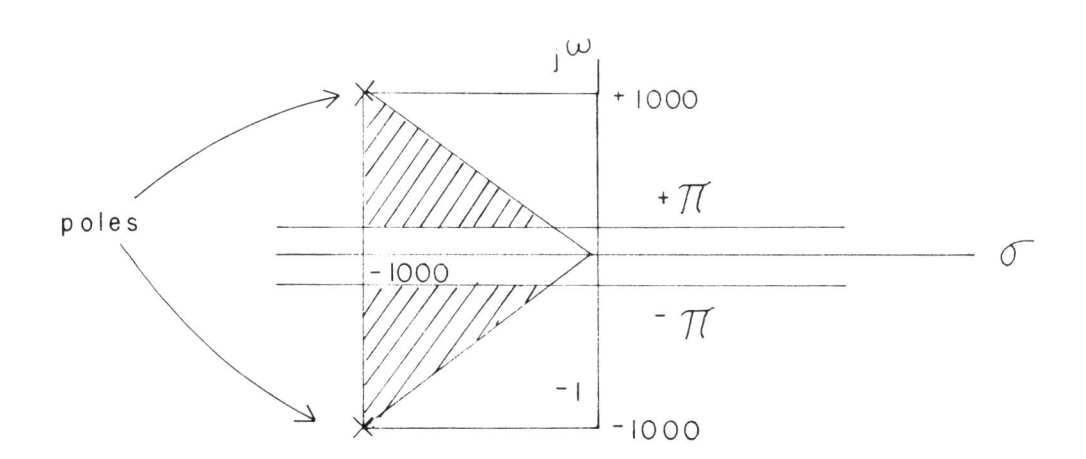

76 / CONTROLS

Problem 2

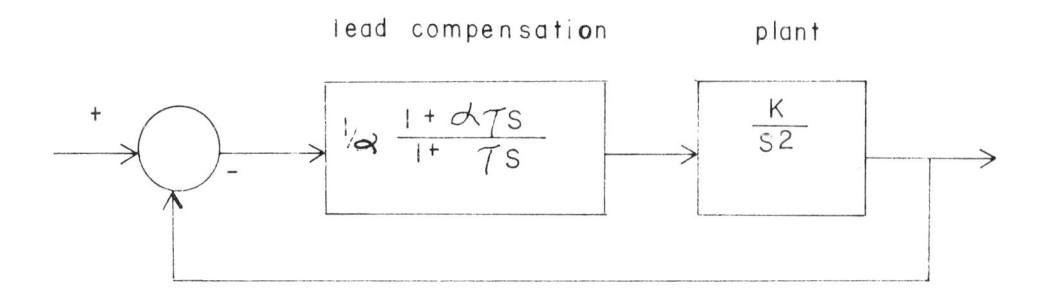

lead compensation — $\frac{1}{\alpha} \frac{1+\alpha Ts}{1+Ts}$

plant — $\frac{K}{s^2}$

Lead compensation has been inserted in this feedback system in order to stabilize it.

For the lead compensation network it can be shown that:

$$\sin \phi_m = \frac{\alpha - 1}{\alpha + 1}, \quad \omega_m = \sqrt{zp}$$

$$z = 1/\alpha\tau \quad p = 1/\tau$$

where ϕ_m is the maximum lead phase angle
ω_m is the frequency at which it occurs
z is the zero of the compensation
p is the pole of the compensation.

Question 2.1: Find α for a $\phi_m = 45°$.

$$\sin \phi_m = \frac{1}{\sqrt{2}} = \frac{\alpha - 1}{\alpha + 1} = .707$$

$$\alpha - 1 = .707\alpha + .707$$

$$.293\alpha = 1.707$$

$$\alpha = 5.83$$

Question 2.2: For $\alpha = 6$ and $\omega_m = 5$ rad/sec, find the pole and zero frequencies, z, p, for the compensation network.

$$z = 1/\alpha\tau \qquad p = 1/\tau$$

$$\omega_m = \sqrt{zp} = \sqrt{1/\alpha\tau^2}$$

$$\tau = 1/\omega_m \cdot 1/\sqrt{\alpha} = 1/12.25$$

Putting the numbers into the expressions for z, p:

$$z = 2.04 \text{ rad/sec} \qquad p = 12.25 \text{ rad/sec}$$

Question 2.3: Find the value of K for a cross-over frequency of 5 rad/sec.

Write the open loop transfer function:

$$H(s) = \frac{K}{\alpha} \frac{1}{s^2} \frac{1 + \alpha\tau s}{1 + \tau s}$$

Take the magnitude and set equal to 1 to find the cross-over frequency:

$$|H(s)| = \frac{K}{6} \frac{1}{25} \frac{\sqrt{1 + (5/2)^2}}{\sqrt{1 + (5/12.2)^2}} = \frac{K}{150} \frac{2.7}{1.08} = 1$$

$$K = 60$$

78 / CONTROLS

Question 2.4: Sketch the magnitude of the open loop transfer function by using the Bode asymptotic method:

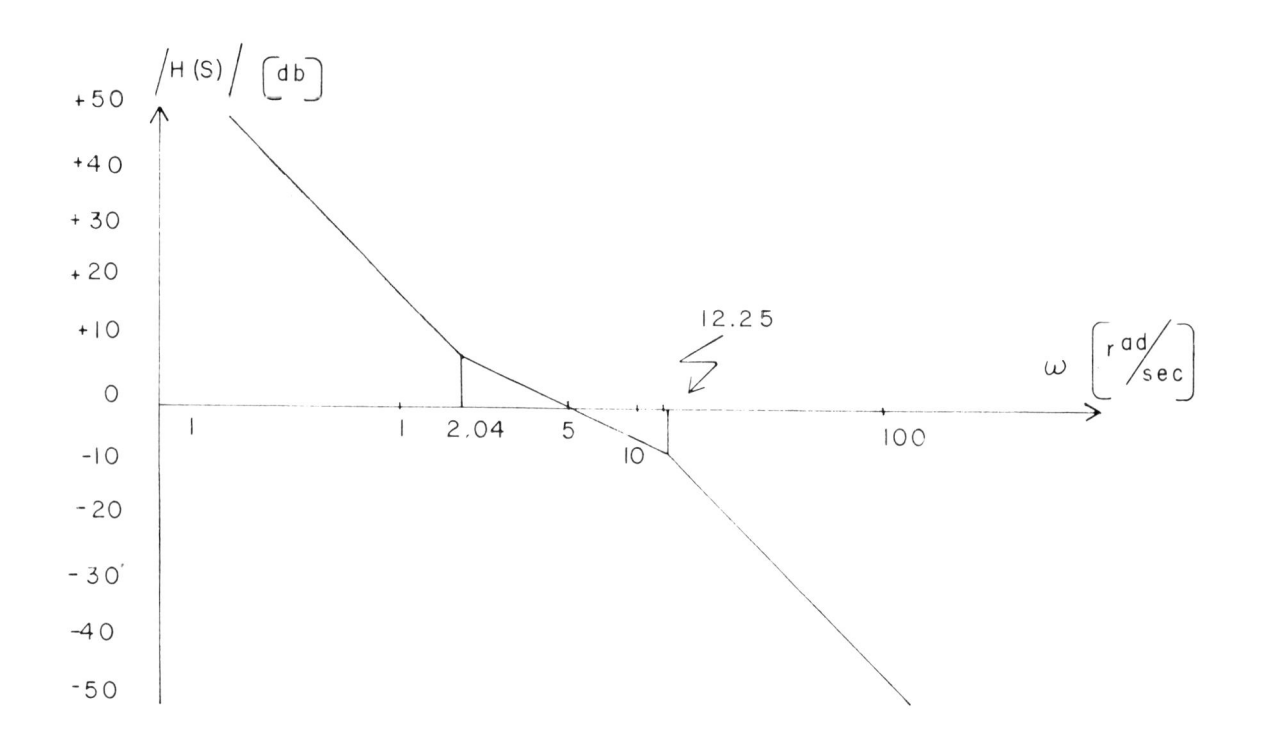

Problem 3

For the following feedback system,

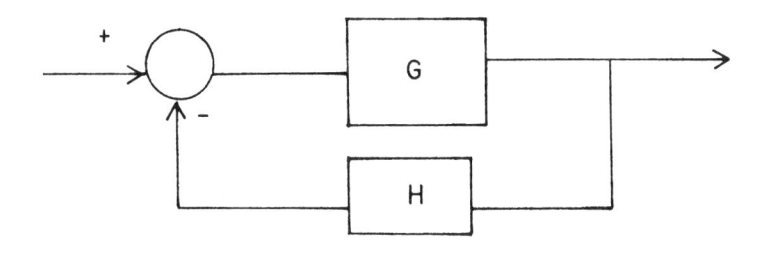

the loop transfer function is:

$$GH = \frac{K(s^2 + 4s + 8)}{s^2(s + 4)}$$

Plot a root locus of GH as the gain, $0 < K < \infty$, by the following steps:

Question 3.1: Plot the open loop poles and zeros in the s-plane.

Do this by first factoring the open loop function:

$$GH = \frac{K(s + 2 + j2)(s + 2 - j2)}{s^2(s + 4)}$$

80 / CONTROLS

Then plot the poles appearing in the denominator and the zeros appearing in the numerator of GH.

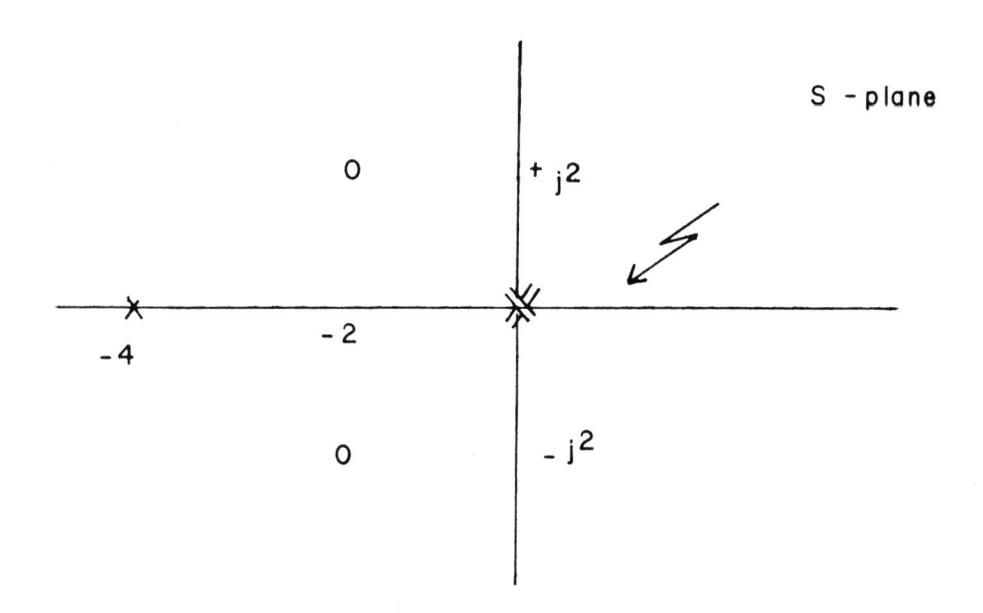

Question 3.2: Locate the segments of the real axis which are root loci.

They are the segments beyond an odd number of poles and zeros.

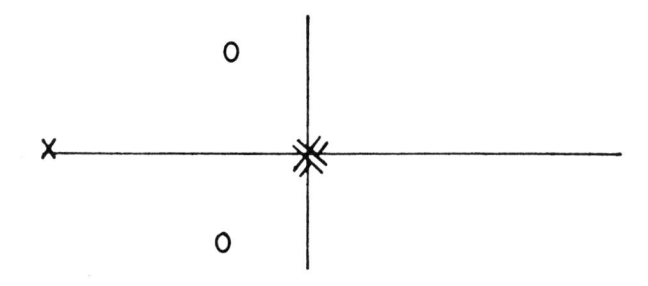

Question 3.3: Determine the number of separate loci.

The number of separate loci is equal to the number of poles.

$$\text{number of loci} = 3$$

Question 3.4: Locate the angles of the asymptotes of the loci and their centroid.

An expression for the angles is:

$$\text{angles} = \frac{2q + 1}{\text{\# poles} - \text{\# zeros}}$$

where $q = 0, 1, \ldots$ (# poles − # zeros − 1).

Therefore,

$$q = 0$$

and

$$\text{angle} = 180°$$

An expression for the centroid is:

$$\text{centroid} = \frac{\Sigma \text{ pole positions} - \Sigma \text{ zero positions}}{\text{\# poles} - \text{\# zeros}}$$

$$\text{centroid} = \frac{4 - 4}{1} = 0$$

82 / CONTROLS

<u>Question 3.5</u>: Estimate the angles of departure at the complex poles and the angles of arrival at the complex zeros.

For the angles of departure, θ_D, draw a root near one of the poles. Then use the following expression:

$$\Sigma \text{ angles from zeros to root} - \Sigma \text{ angles from poles to root} = 180°$$

Doing this:

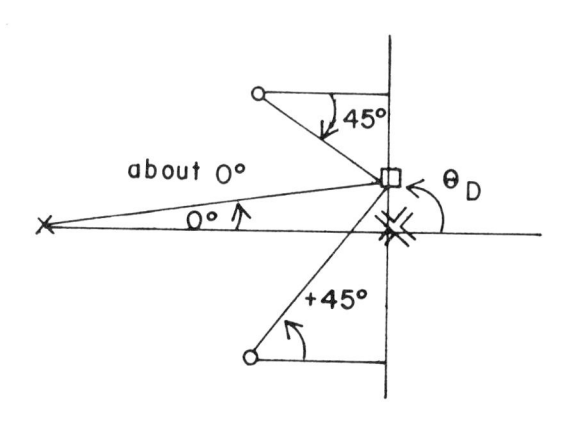

[Diagram showing complex plane with poles and zeros, angles of 45° at top, about 0°, 0°, and +45° at bottom, with θ_D indicated]

$$(+45° - 45°) - (0° + 2\theta_D) = 180°$$

$$\theta_D = 90° \quad \text{angle of departure}$$

For the angles of arrival, θ_A, draw a root near one of the zeros. Then use the following expression:

$$\Sigma \text{ angles from zeros to root} - \Sigma \text{ angles from poles to root} = 180°$$

Doing this:

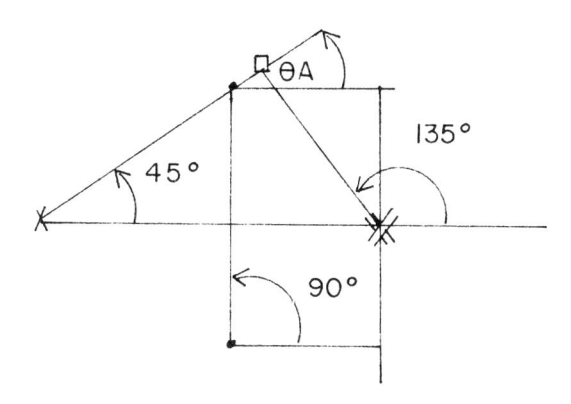

$$(\theta_A + 90°) - (45° + 2(135°)) = 180°$$

$$\theta_A = 45° \quad \text{angle of arrival}$$

Question 3.6: Check to see if the complex poles ever cross the imaginary axis by using a Routh-Hurwitz array.

Write down the characteristic equation, which is the denominator of the closed loop transfer function.

$$1 + \frac{K(s^2 + 4s + 8)}{s^2(s + 4)} = 0$$

or

$$s^3 + (4 + K)s^2 + 4Ks + 8K = 0$$

Then, the following array is made up:

$$
\begin{array}{c|cc}
s^3 & 1 & 4K \\
s^2 & 4+K & 8K \\
s^1 & \dfrac{4K(2+K)}{4+K} & 0 \\
s^0 & 8K & 0
\end{array}
\equiv
\begin{array}{c|cc}
s^3 & a_3 & a_1 \\
s^2 & a_2 & a_0 \\
s^1 & b_2 & b_1 \\
s^0 & c_2 & c_1
\end{array}
$$

The first row is coefficients of odd powers of s.
The second row is coefficients of even powers of s.
The third row is

$$b_2 = -\frac{a_3 a_0 - a_2 a_1}{a_2}, \quad b_1 = 0$$

The fourth row is

$$c_2 = -\frac{a_2 b_1 - a_0 b_2}{b_2}, \quad c_1 = 0$$

The number of changes of sign in the first column of the Routh-Hurwitz array indicates the number of poles in the right half s-plane. Since for $0 < K < \infty$ there are no changes of sign, there are no complex poles which cross the imaginary axis.

Question 3.7: Draw the root locus.

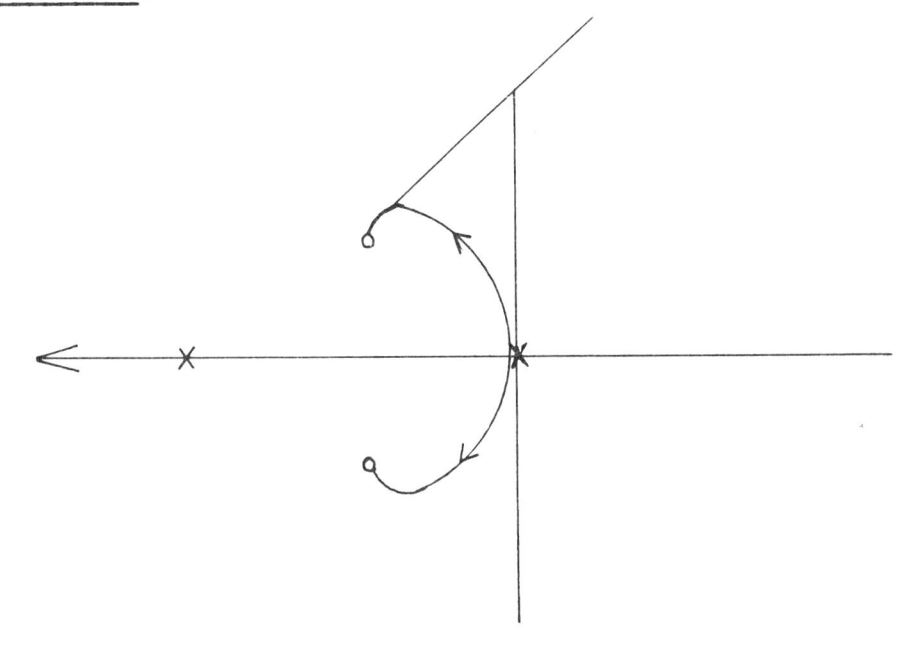

Problem 4

Given the following feedback system

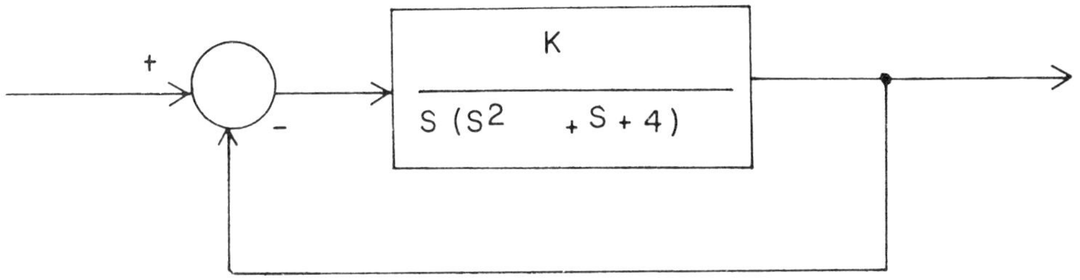

Question 4.1: Find out if the system is stable for a gain of 8 (K=8), by using the Routh-Hurwitz array.

Write down the characteristic equation:

$$1 + \frac{8}{s(s^2 + s + 4)} = 0$$

or

$$s^3 + s^2 + 4s + 8 = 0$$

Make a Routh-Hurwitz array.

s^3	1	4
s^2	1	8
s^1	-4	0
s^0	8	0

Since there are two changes of sign in the first column, there are two poles in the right half s-plane. The system is unstable.

86 / CONTROLS

Question 4.2: Make a Nyquist plot for the system.

Write down the forward transfer function:

$$G(s) = u + jv = \frac{K}{s(s^2 + s + 4)}$$

Make a table of magnitude of $G(s)$, angle of $G(s)$, for different frequencies, ω.

$|G(s)|$	$\angle G(s)$	ω
∞	$-90°$	0
0	$-270°$	∞
$-K/4$	$-180°$	2

The last table entry is found by the following:

$$G(s) = \frac{-K\omega^2}{\omega^4 + (4\omega - \omega^3)^2} + j \frac{-K(4\omega - \omega^3)}{\omega^4 + (4\omega - \omega^3)^2}$$

The $I_m G(s) = 0$ when $4\omega - \omega^7 = 0$ or $\omega = 2$

Then $ReG(s) = -K/4$ and $\angle G(s) = 180°$

Plotting this information as a magnitude, angle polar plot:

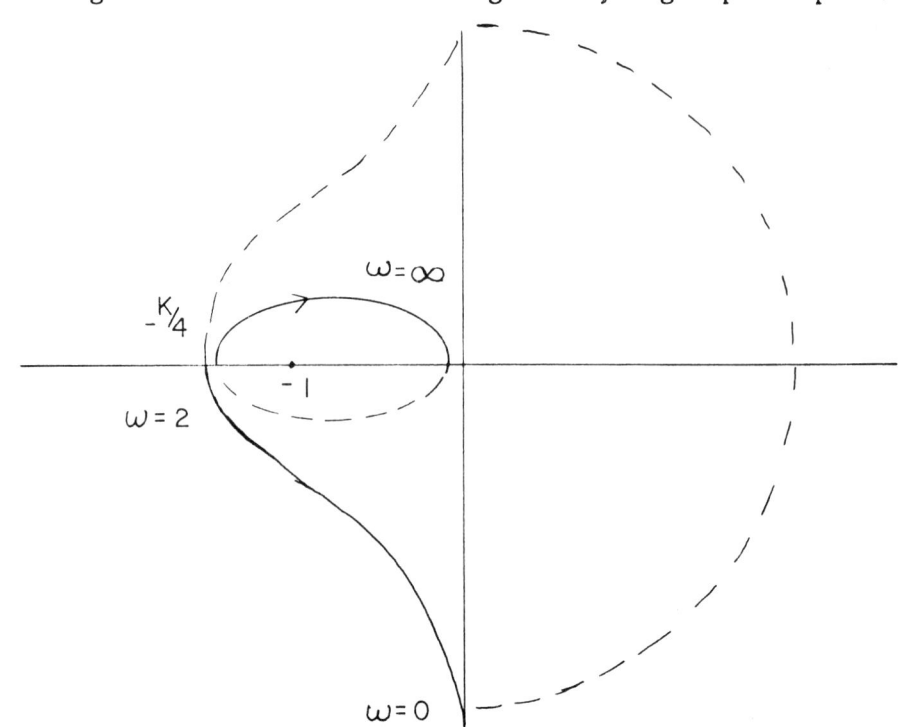

Question 4.3: State the Nyquist stability criteria.

$$\begin{array}{c}\text{the number of clockwise}\\ \text{encirclements of -1 point}\end{array} - \begin{array}{c}\text{the number of counter-}\\ \text{clockwise encirclements}\end{array} = \begin{array}{c}\text{number of right}\\ \text{half plane poles}\end{array}$$

Question 4.4: For what values of K is the system stable.

If the magnitude of G(s) is less than -1 at an angle of 180°, then there are no encirclements of the -1 point and hence the system is stable.

Imposing this condition:

$$-\frac{K}{4} < -1 \quad \text{or} \quad K < 4 \quad \text{for a stable system}$$

Question 4.5: How many right half plane poles are there for a K = 16.

There are two.

References:

Brogan, William L.; <u>Modern Control Theory</u>, Quantum Publishers, Incorporated, 1974. See chapter 2.

DiStefano, Stubberud, Williams; <u>Feedback and Control Systems</u>, Schaum's Outline Series, 1967. See chapters 3,5,6,7,9,10,11,13,15.

Chapter 7

COMMUNICATIONS

Problem 1: The following diagram is of a stereophonic transmitter:

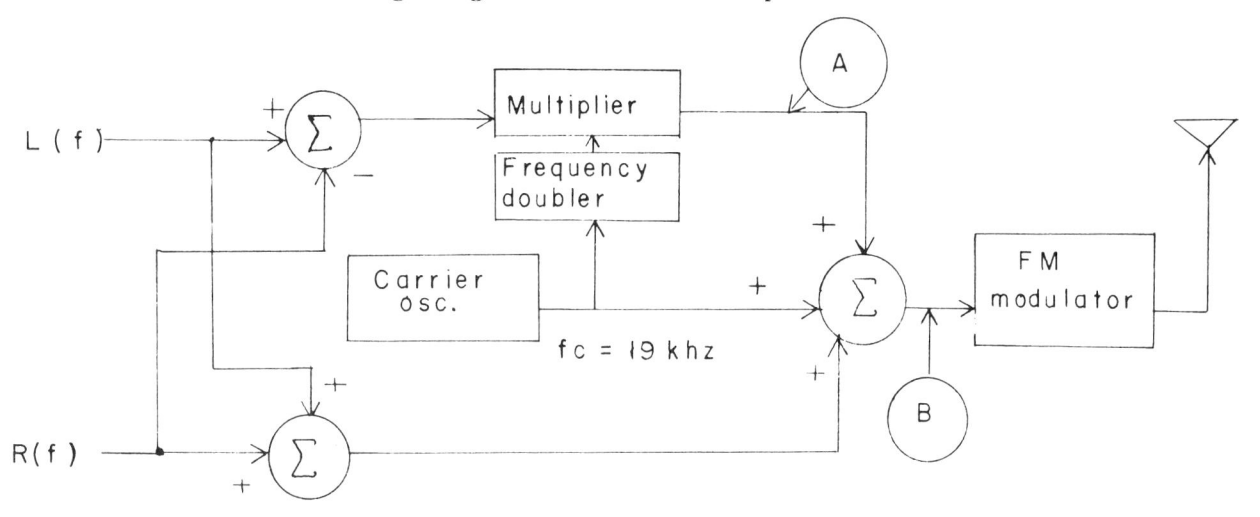

where L(t) is the left channel signal and R(t) is the right channel signal

Magnitudes of the frequency spectra for the left channel and the right channel are shown below:

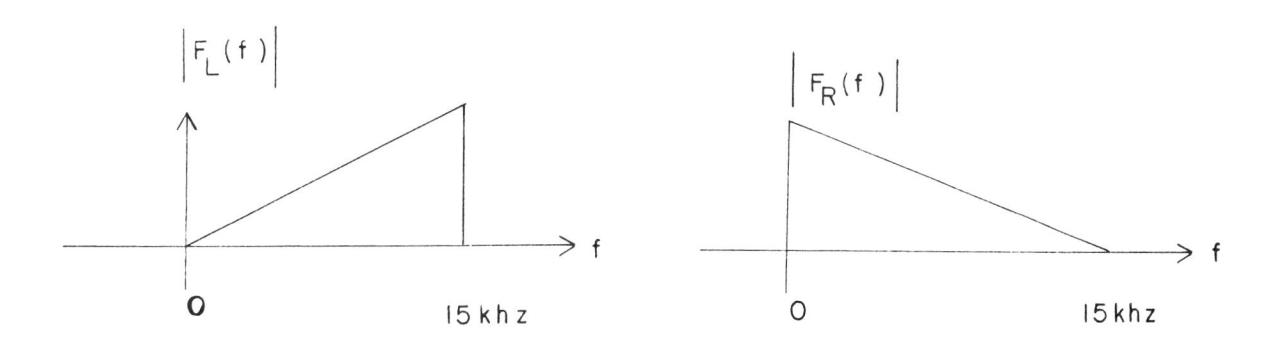

89

Question 1.1 Draw the magnitude of the frequency spectra to be seen at points A and B, assuming that all frequency components in A and B are in phase.

The magnitude of the spectra L - R is:

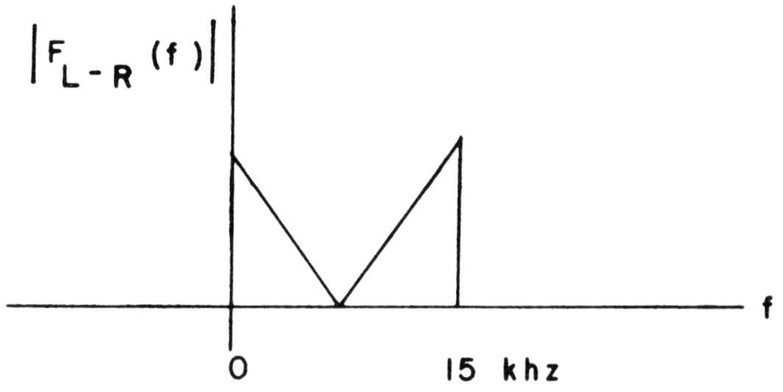

|F L-R (f)|
f
0
15 khz

This signal is then modulated by 38 khz. So the components are $2f_c + f_s$ and $2f_c - f_s$.

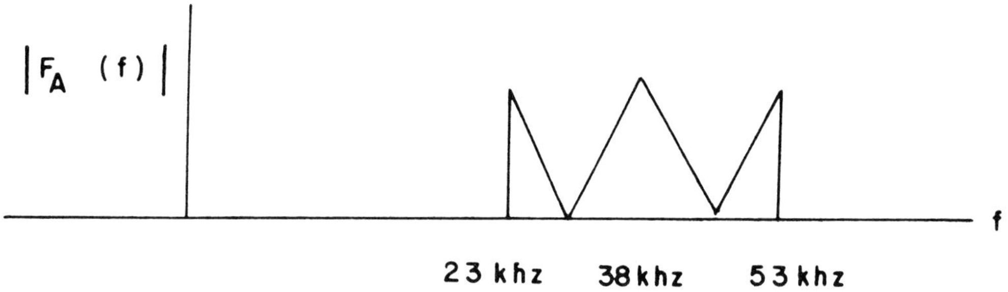

|F A (f)|
f
23 khz
38khz
53khz

The magnitude of the spectra L + R is:

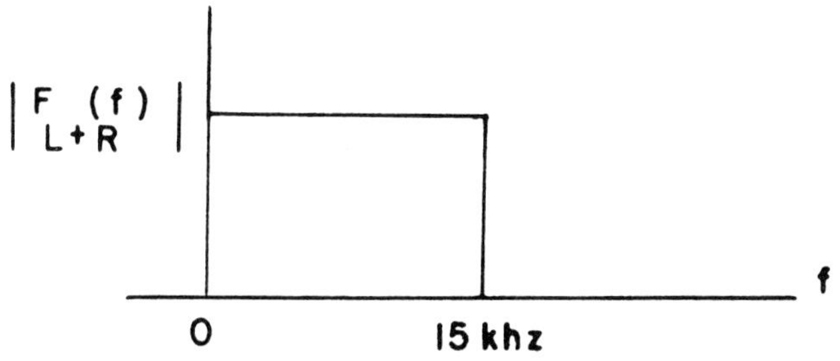

|F L+R (f)|
f
0
15khz

The three signals f_c, L + R, L − R, are summed at point B. This gives the magnitude frequency spectra shown:

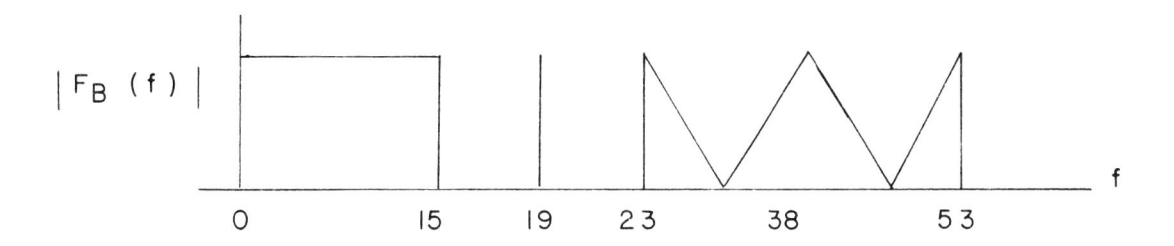

[Spectrum |F_B(f)|: rectangular block from 0 to 15, impulse at 19, triangular spectra centered around 23–38–53, on f axis]

Question 1.2 The listener who has a monaural receiver has the following:

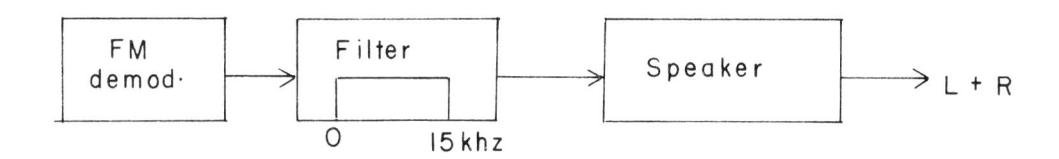

[Block diagram: FM demod. → Filter (0 to 15 kHz) → Speaker → L + R]

Show the magnitude of the frequency spectra which he listens to.

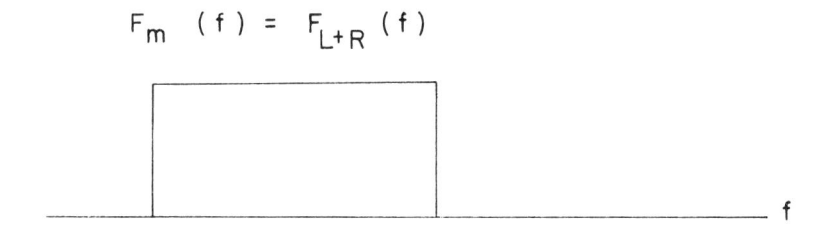

$$F_m(f) = F_{L+R}(f)$$

[Rectangular spectrum on f axis]

Question 1.3 The stereophonic receiver is shown in the following block diagram.

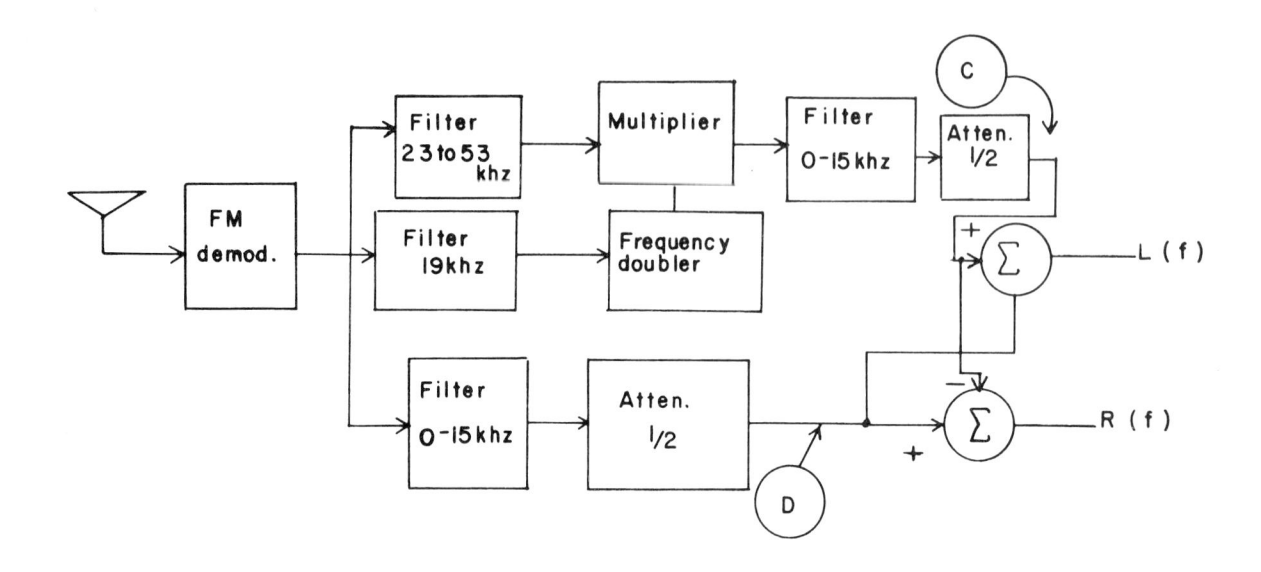

Draw the magnitude of the frequency spectra at points C and D.

The signal between 23 khz and 53 khz is demodulated by 38 khz and then filtered by the low pass, 0 - 15 khz filter. The components of the signal are then $f_s - 2f_c$.

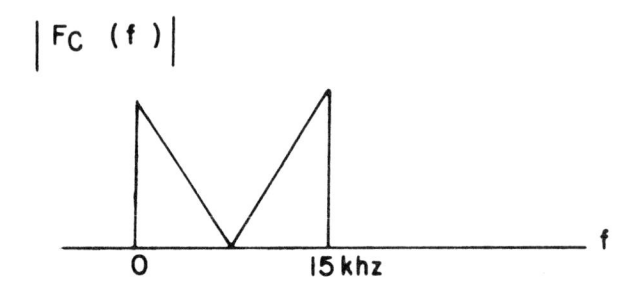

The signal at point D is that which passes through a 0 - 15 khz low pass filter.

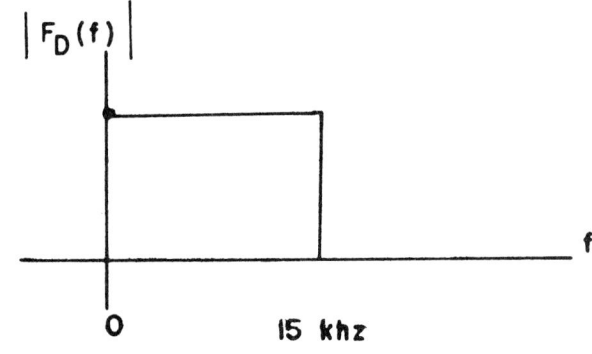

Question 1.4 If there were no monaural listeners, so that L + R was not needed at low end of the frequency spectra, how could the stereo system by simplified.

Simply seperate the left and right frequency spectra into different frequency bands. For the transmitter:

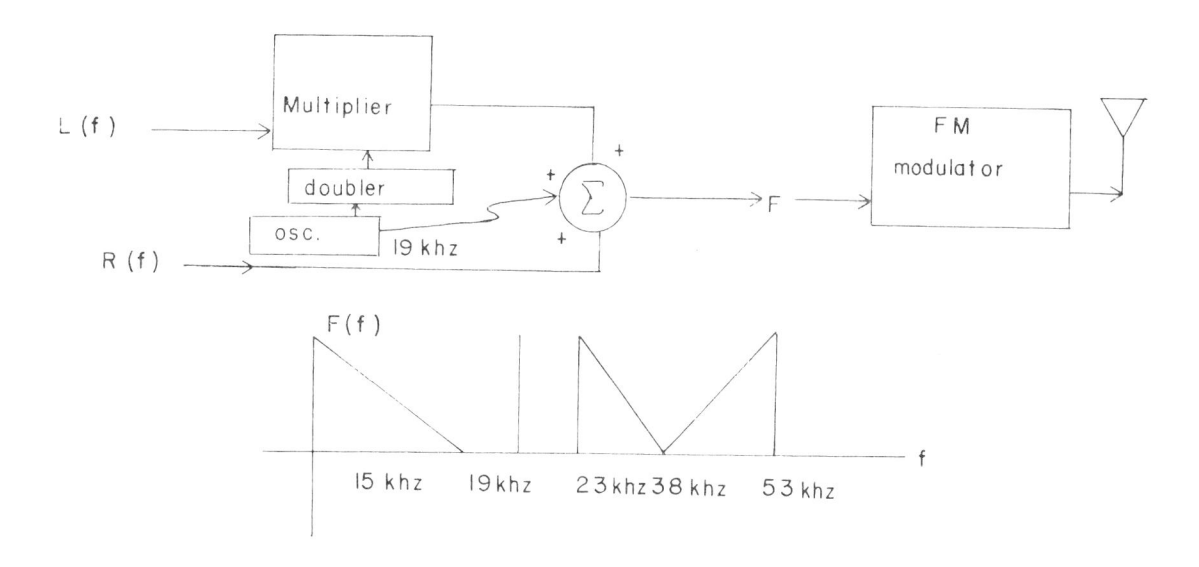

Then, using appropriate filters in the receiver, demodulate the frequency multiplexed signal back into L and R. For the receiver:

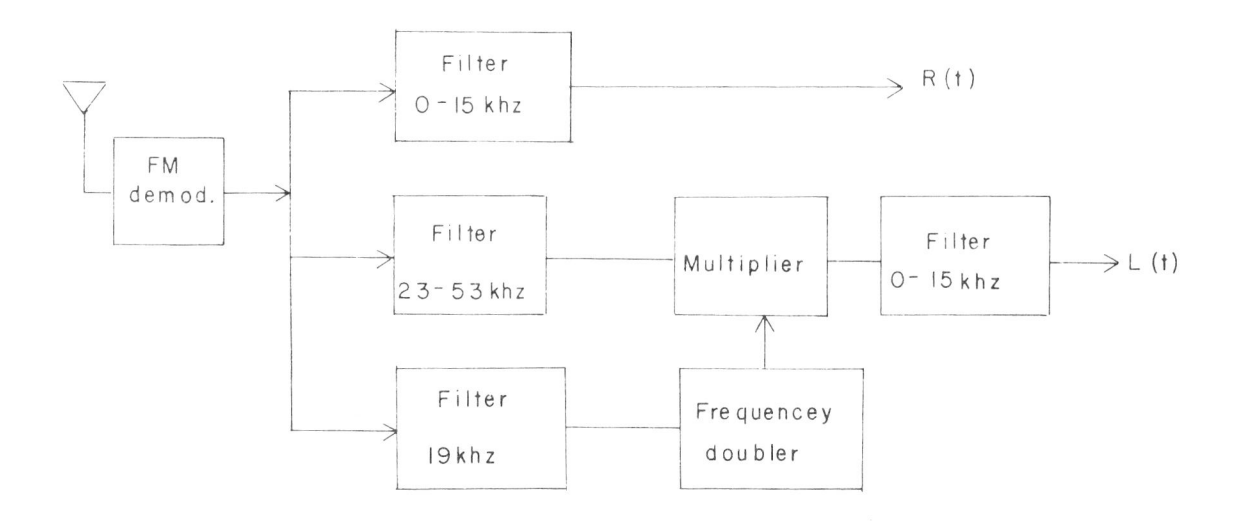

Problem 2: Match a 300Ω antenna to a 50Ω line. This can be done by means of a shorted stub of length, d_2, located at a distance, d_1, from the antenna as shown in the diagram.

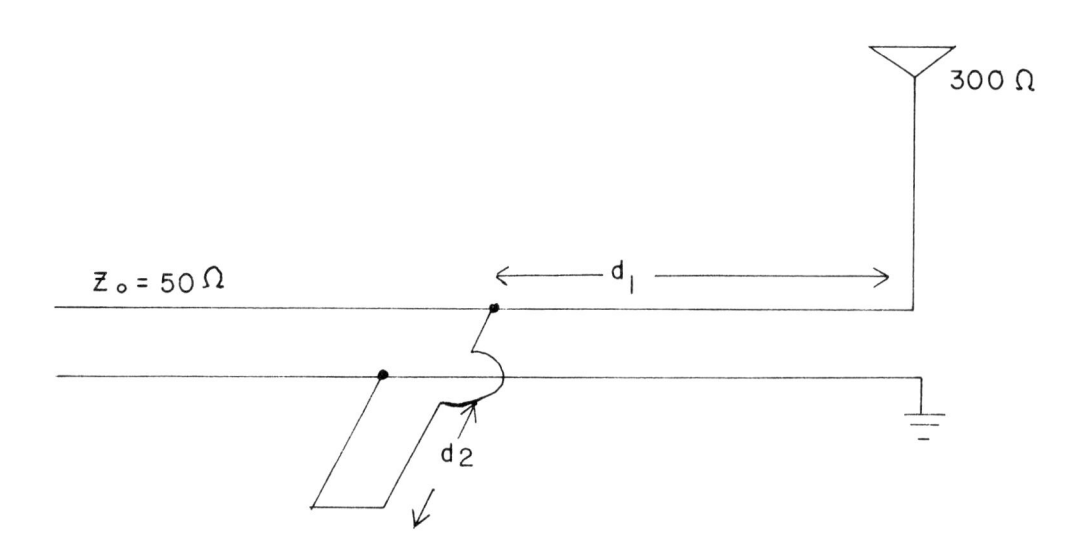

A step-by-step solution by using the Smith Chart is required.

Question 2.1 Find the value of normalized load admittance.

$$Z_{Ln} = \frac{Z_L}{Z_o} = \frac{300}{50} = 6$$

$$Y_{Ln} = \frac{1}{Z_{Ln}} = .16$$

Question 2.2 Enter the value of normalized load admittance on the Smith Chart*.

Y_{Ln} = .16 is entered on the horizontal line through the center of the chart. It is marked as point A.

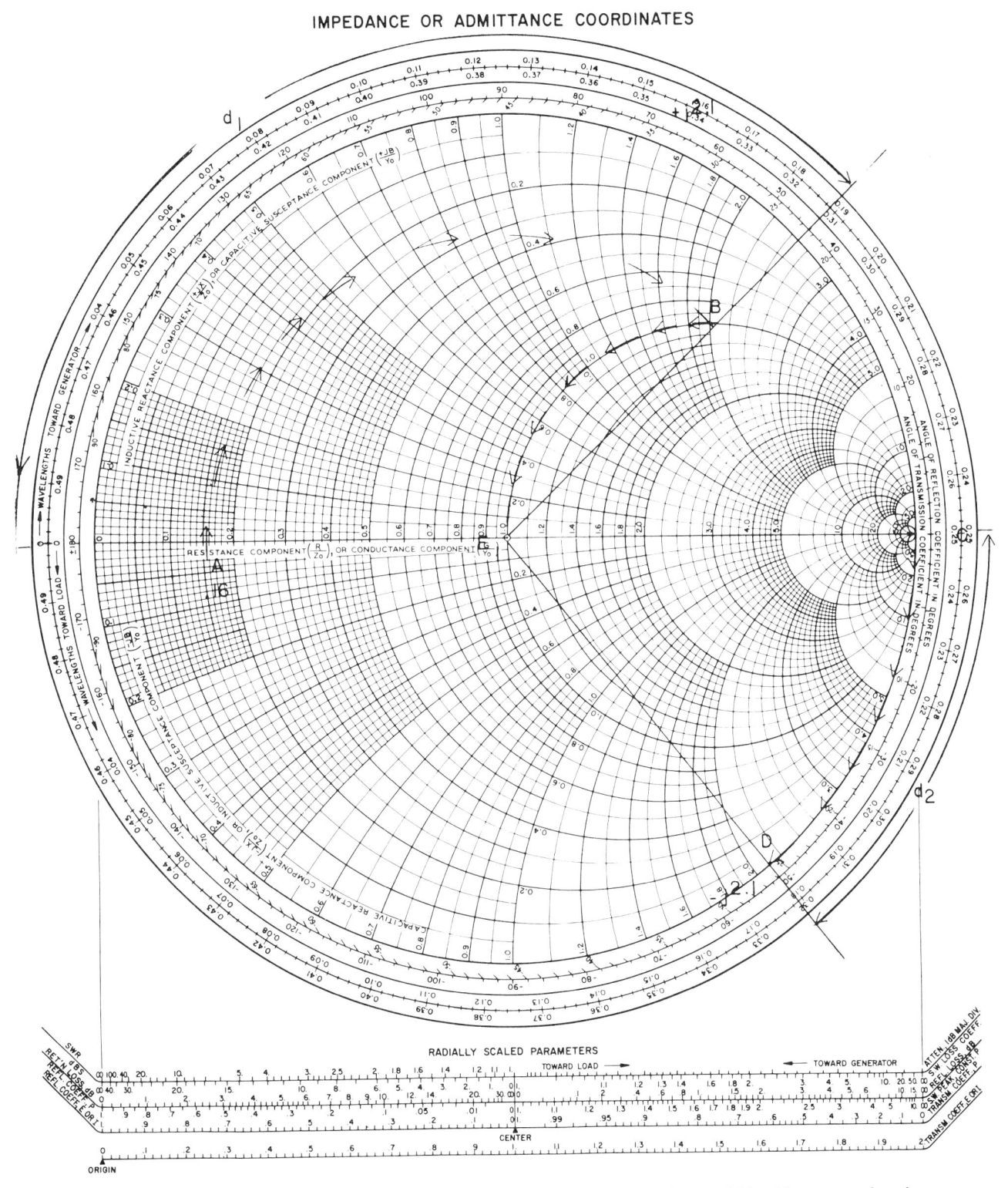

*Smith Chart is copyrighted by, and reproduced herein, with the permission of Kay Elemetrics Corp., Pine Brook, N.J.

Question 2.3 Find the admittance value which the stub must present to the line in order to cancel out the reactive component of the line.

This is done by swinging an arc, with a compass from .16 on the real axis with center at 1 on the real axis, until it intersects the circle which passes through 1 on the Smith Chart. This intersection point is point B.

The reactive component is then read from the chart by following the reactive curve of the Smith Chart, which passes through point B, to the outside circle of the Smith Chart.

The admittance value is +j2.1.

Question 2.4 Find out how many wavelengths d_1 is.

d_1 is read from the Smith Chart as being the length of the arc from A to B. This is read from the outside circle as:

$$d_1 = .120\lambda$$

Question 2.5 Find out how long d_2 is. Its admittance must cancel the reactive component at d_1 on the line.

The stub has an infinite admittance at the shorted end. This is entered on the Smith Chart as point C. An arc is then swung from point C with the center of the chart as its center until -j2.1 is found on the outside circle of the Smith Chart. This is marked as point D.

The distance d_2 is the length of the arc in wavelengths read from the chart.

$$d_2 = (.326 - .250)\lambda = .076\lambda$$

Question 2.6 What happens when the admittance at point B is added to the admittance at point D.

$$Y_{nB} = 1 + j2.1$$

$$Y_{nD} = -j2.1$$

$$Y_{nB} + Y_{nD} = 1 + j2.1 - j2.1 = 1$$

Since the reactive components cancel, the resulting admittance is real and equal to that of the line. Hence the antenna has been matched to the line. This is shown as point E on the Smith Chart.

Problem 3: The following signal in time is applied to an ideal low pass filter as shown. Find the magnitude of the amplitude spectra and sketch it.

COMMUNICATIONS / 97

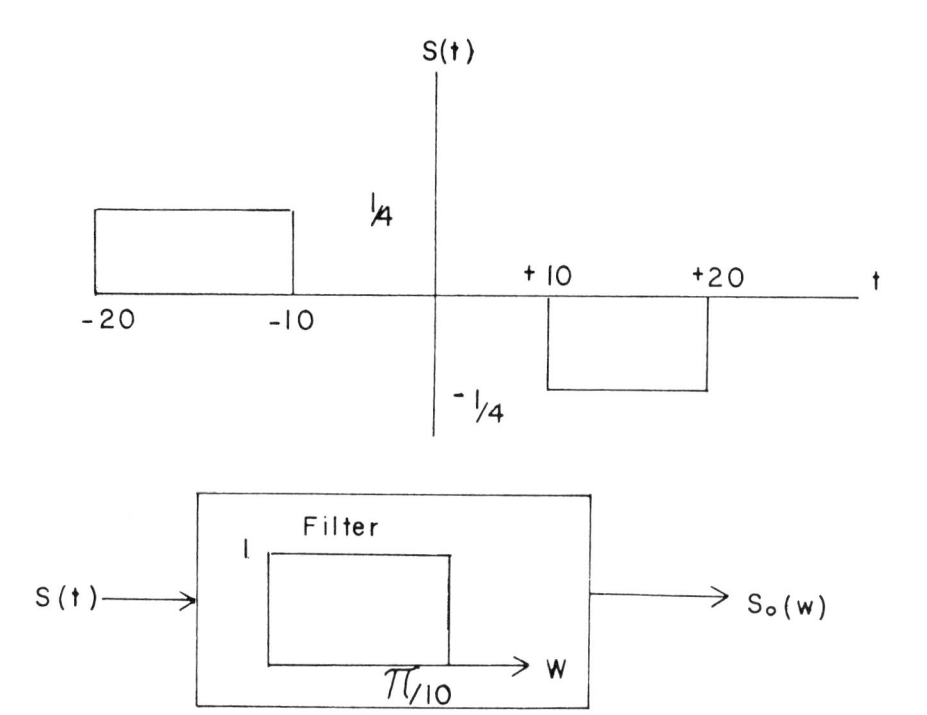

This will be done by the following steps.

Question 3.1 Draw the time derivative of S(t).

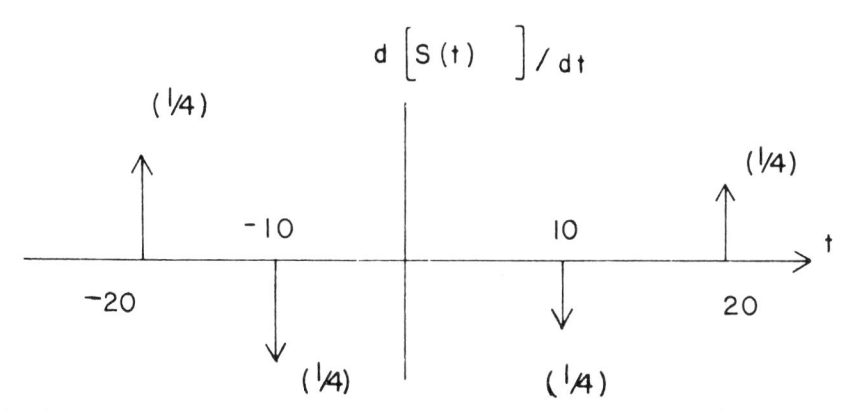

It is well known from theory of singularity functions (i.e., impulses, steps, ramps, etc.) that the derivative of a step is an impulse with an area equal to the amplitude magnitude of the step.

Question 3.2 Transform the signal d s(t)/dt into the frequency domain.

It is known from the theory of singularity functions that an impulse transforms into a constant.

$$\delta(t) \leftrightarrow 1$$

It is further known from the shifting theorem that a time shifted function transforms into the following

$$f(t-T) \leftrightarrow F(\omega) e^{-j\omega T}$$

Using these two theorems, the transform of $s'(t)$ is:

$$F'(\omega) = \tfrac{1}{4} \; e^{+j20\omega} - e^{j10\omega} - e^{-j10\omega} + e^{-j20\omega}$$

$$F'(\omega) = \tfrac{1}{2} (\cos 20\omega - \cos 10\omega)$$

Question 3.3   Sketch $S'(\omega)$

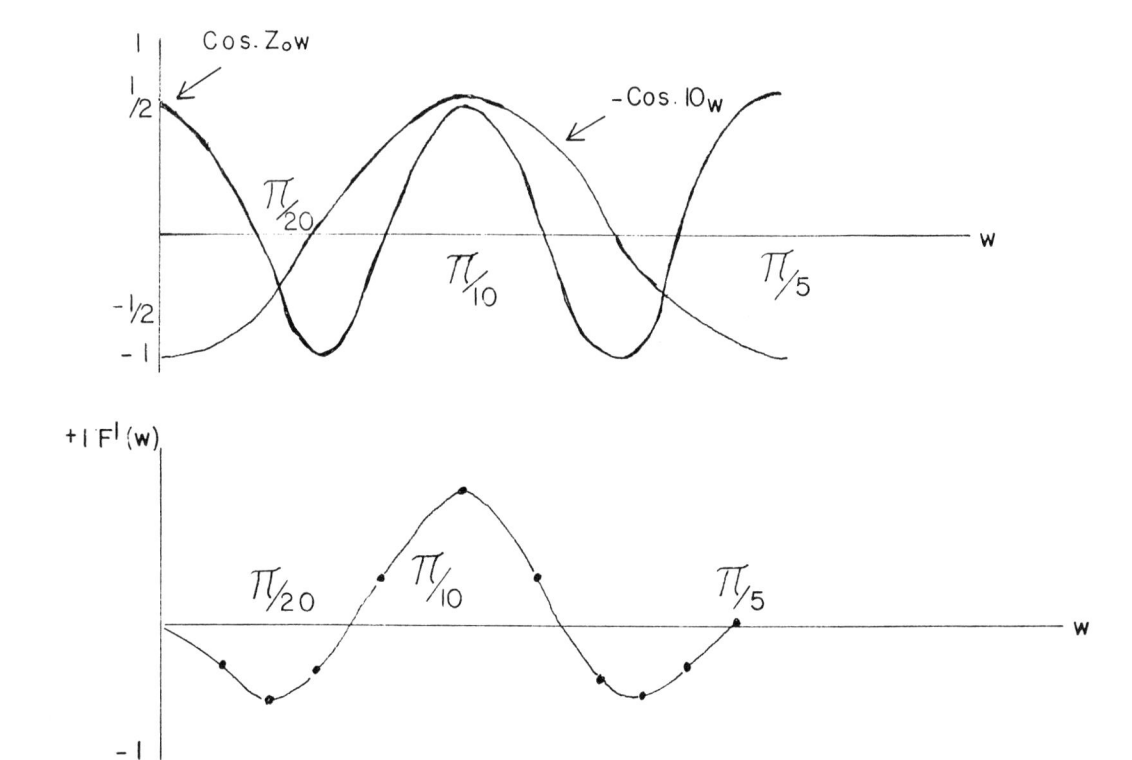

This is simply an exercise in adding two sinusoids of different frequency which results in a non-sinusiodal wave form as shown above.

Question 3.4   Find the analytic expression for the transform of the original $s(t)$.

By using the transform theory integral theorem:

$$\int f(t) \, dt \leftrightarrow \frac{F(\omega)}{j\omega}$$

Therefore,

$$S(\omega) = \frac{S'(\omega)}{j\omega}$$

$$S(\omega) + \frac{\frac{1}{2}(\cos 20\omega - \cos 10\omega)}{j\omega}$$

Question 3.5 Sketch the amplitude magnitude of $S(\omega) = |S(\omega)|$.

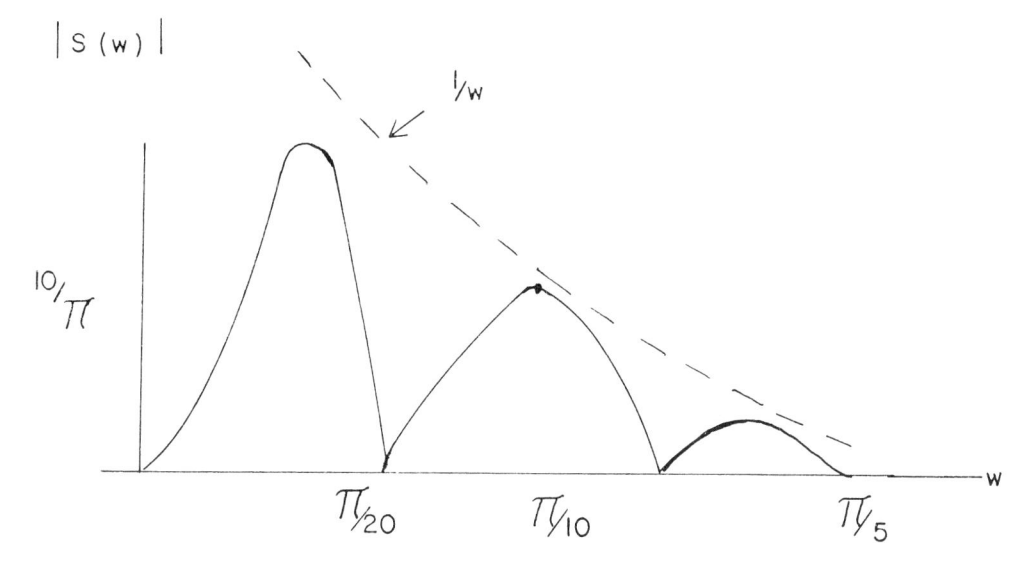

[Sketch: $|S(\omega)|$ vs ω, with peak value $10/\pi$, lobes at $\pi/20$, $\pi/10$, $\pi/5$, enveloped by $1/\omega$ dashed curve]

Question 3.6 Find the magnitude of the output amplitude spectra, $|S_o(\omega)|$.
This is done by using the well known theorem of system theory:

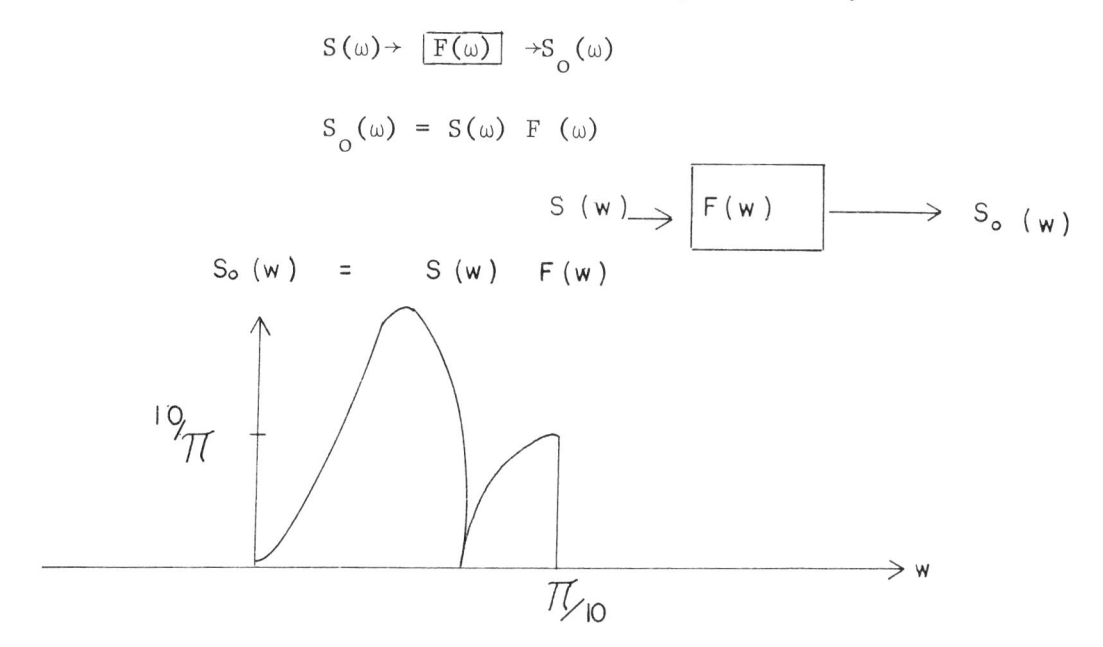

$$S(\omega) \rightarrow \boxed{F(\omega)} \rightarrow S_o(\omega)$$

$$S_o(\omega) = S(\omega) F(\omega)$$

$S(\omega) \rightarrow \boxed{F(\omega)} \rightarrow S_o(\omega)$

$S_o(\omega) = S(\omega) \; F(\omega)$

[Sketch: $S_o(\omega)$ with amplitude $10/\pi$, extending to $\pi/10$]

Problem 4: The following is a digital transmission system for a voice signal

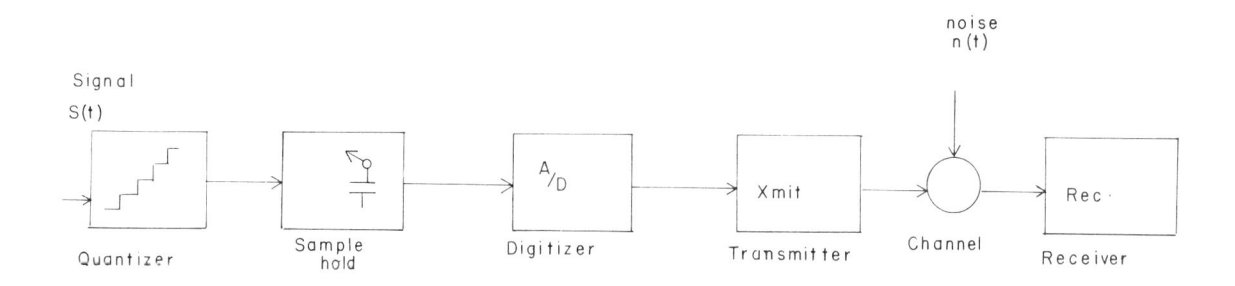

The voice signal, s(t), is quantized, sampled, digitized and then sent over a channel where Gaussian noise (n(t)) is added. The signal then goes to the receiver. The channel bandwidth is limited to 4 khz.

Question 4.1 If the quantizer has 16 levels and its output is sampled at an 8 khz rate, find the information rate out of the transmitter.

The information rate, I, is equal to the sample rate, S, times the log to the base 2 of the number of quantization levels, M.

$$I = S \log_2 M$$
$$I = (8) \log_2 16 \text{ khz} = 32 \text{ khz.}$$

Question 4.2 Find the signal to noise ratio, S/N, for the channel.

From Shannon's theorem:

$$I = W \log_2 (1 + S/N)$$

where W is the channel bandwidth.

By solving for S/N:

$$S/N = 2^{I/W} - 1$$
$$= 2^8 - 1 = 256 - 1$$
$$= 255$$

Question 4.3 If the transmitter power were increased so that S/N = 1023, what could the channel bandwidth be reduced to?

$$W = \frac{I}{\log_2 1024} = \frac{32 \text{ khz}}{10}$$

$$W = 3.2 \text{ khz}$$

Question 4.4 If the transmitter power is set so that S/N = 1023 and the channel bandwidth stays at 4khz, what can the information rate be increased to?

$$I = W \log_2(1 + S/N)$$

$$I = (4\text{khz}) \log_2 1024$$

$$I = 40 \text{ khz.}$$

Question 4.5 If the same conditions in question 4.4 hold and the quantization stays the same, what could the sampling rate be increased to?

$$I = S \log_2 M$$

$$S = \frac{I}{\log_2 M} = \frac{40\text{khz}}{4}$$

$$S = 10\text{khz.}$$

References:

Hancock, J. C., <u>Introduction to Electrical Design</u>, Holt, Rinehart, and Winston, Inc. 1972.

Schwartz, Mischa, <u>Information Transmission, Modulation, and Noise</u>, 2nd Edition, McGraw-Hill, 1970.

Chapter 8

ELECTRONICS

Problem 1

This is a difference amplifier.

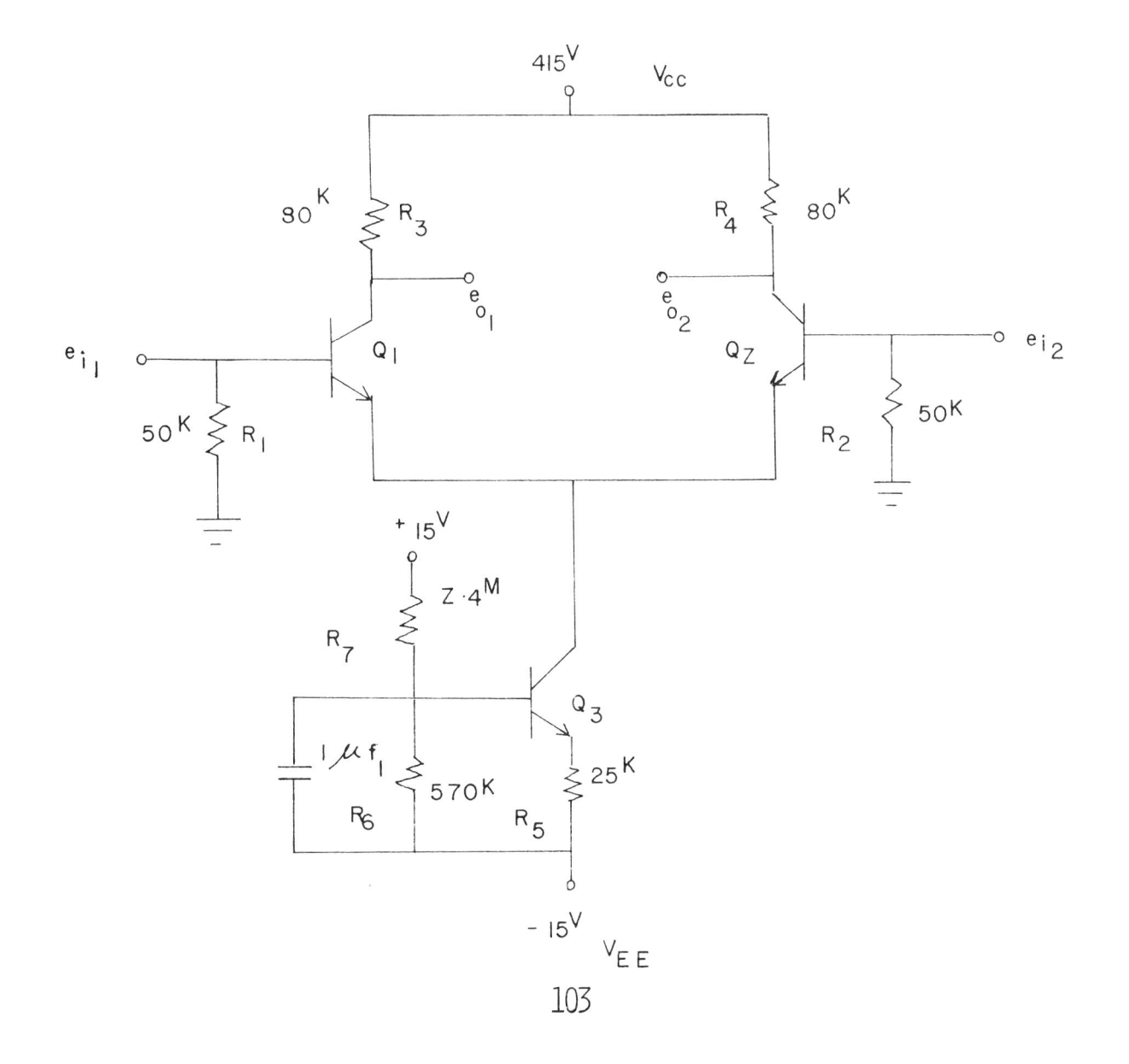

In this configuration, transistor Q3 acts like a constant current source or an equivalent very large resistor. The equivalent value will be found in order to compute such properties as sum gain, difference gain, common mode, rejection ratio, etc., of the amplifier.

Question 1.1: For zero volts on both inputs, find the quiescent or D.C. operating point for transistor Q3 (i.e., find V_{CE} and I_C).

The emitter voltage, V_E, is:

$$V_E = \frac{R_6}{R_6 + R_7} (V_{CC} - V_{EE}) - V_{BE} = 5 \text{ v.}$$

The collector current, I_C, is:

$$I_C = \frac{V_E}{R_5} = 200 \text{ μa}$$

The collector voltage, V_C, is:

$$V_C = -V_{BE} = -.750 \text{ v.}$$

The collector to emitter voltage, V_{CE}, is:

$$V_{CE} = -V_{BE} - V_{EE} - V_E = 9.25 \text{ v.}$$

$$V_{CE} = 9.25 \text{ v.} \qquad I_C = 200 \text{ μa}$$

Question 1.2: Find the transconductance, g_m, and the input resistance, r_π, of the small signal hybrid π model for transistor Q3.

$$g_m = \frac{qI_C}{KT} = .008 \text{ mho}$$

$$r_\pi = \frac{\beta}{g_m} = 43.7 \text{ K}\Omega$$

Question 1.3: Find the effective resistance seen looking into the collector of transistor Q3.

Do this by first drawing the small signal circuit.

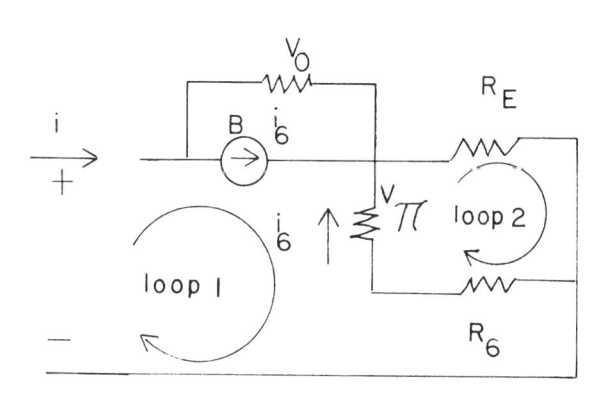

Then write two Kirchoff Voltage Law loop equations:

$$-v + (i - \beta i_b)r_o - i_b(r_\pi + R_b) = 0$$

$$(i + i_b)R_E + i_b(r_\pi + R_b) = 0$$

⎫ 2 loop
⎬ equation
⎭

By collecting terms, the equations become:

$$ir_o - (\beta r_o + r_\pi - R_b)i_b = v$$

$$iR_E + (R_E + r_\pi + R_b)i_b = 0$$

⎫ collecting
⎬ terms
⎭

By using Cramer's Rule, we can solve for the current i:

$$i = \frac{\begin{vmatrix} v & -(\beta r_o + r_\pi + R_b) \\ 0 & R_E + r_\pi + R_b \end{vmatrix}}{\begin{vmatrix} r_o & -(\beta r_o + r_\pi + R_b) \\ R_E & R_E + r_\pi + R_b \end{vmatrix}}$$

$$= \frac{(R_E + r_\pi + R_o)v}{r_o R_E + r_o h_{ie} + v_o R_b + \beta r_o R_E + R_E r_\pi + R_E R_b}$$

The effective resistance is the ratio of v/i:

$$R_e = \frac{v}{i} = \frac{(\beta + 1)r_o R_E + (R_E + r_o)(r_\pi + R_b)}{R_E + r_\pi + R_b} = 28 \text{ M}\Omega$$

<u>Question 1.4</u>: Find the quiescent or D.C. operating point for transistors Q1 and Q2.

$$I_{C1} = I_{C2} = \frac{I_{C3}}{2} = 100 \text{ }\mu\text{a}$$

$$V_{CE1} = V_{CE2} = V_{CC} + V_{BE} - I_{C1}R_3 = 7.75 \text{ v.}$$

<u>Question 1.5</u>: Find the g_m and r_π for transistors Q1 and Q2.

$$g_m = \frac{qI_{C1}}{KT} + .004 \text{ mho}$$

$$r_\pi = \frac{\beta}{g_m} = 87.4 \text{ K}\Omega$$

Question 1.6: Find the common mode gain. This is done by making $e_{i1} = e_{i2}$. Since the voltages on either side of a vertical line splitting the amplifier in half are equal, no current flows from Q1 to Q2. Therefore, the small signal circuit becomes

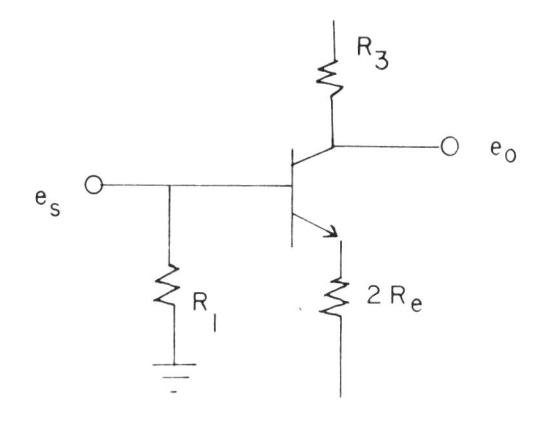

and the small signal model is:

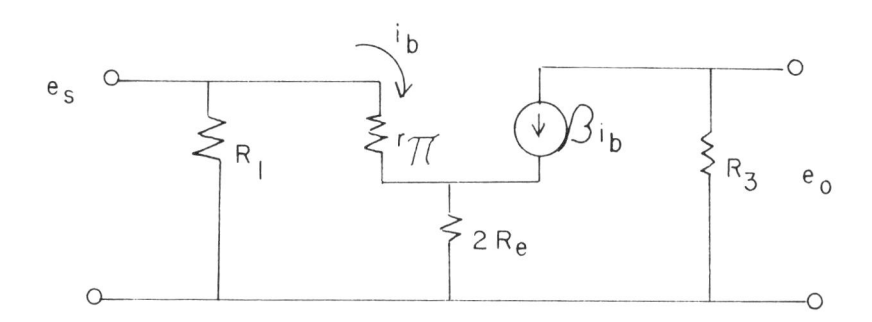

The common mode gain is:

$$A_C = \frac{e_o}{e_s} = \frac{-\beta R_o R_1}{R_1 + r_\pi + 2(\beta + 1)R_E} = -71.5$$

Question 1.7: Find the difference mode gain. This is done by making $e_{i1} = -e_{i2}$. Since the voltages on either side of a vertical line splitting the amplifier in half are equal and opposite in sign, all the current from Q1 flows into Q2. Therefore, the small signal circuit becomes:

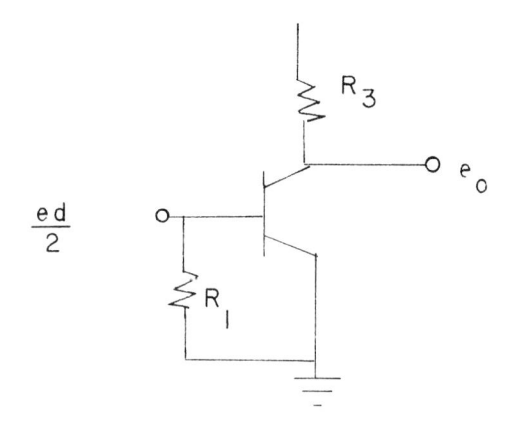

and the small signal model is:

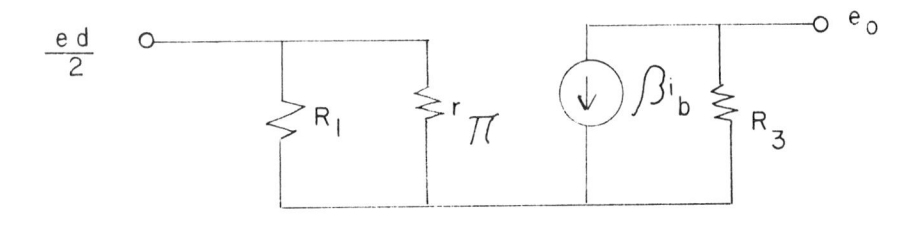

$$A_d = -\frac{R_1 \beta R_3}{R_1 + r_\pi} = -1.42 \times 10^6$$

Question 1.8: Find the common mode rejection ratio.

$$\text{CMRR} \triangleq \frac{A_d}{A_c} = 1.99 \times 10^4$$

Problem 2

This is a two-stage capacitor coupled transistor amplifier.

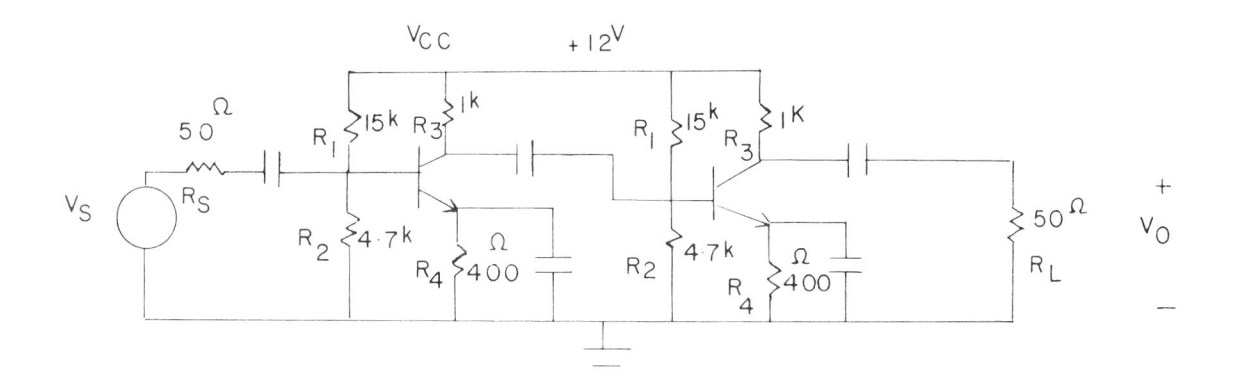

$V_{be} = 750$ mv; $\beta = 80$; $r_o = 12$ k; $r_\pi = 400\ \Omega$; $r_x = 50\ \Omega$; $r_\mu = 1$ MΩ; $C_\mu = 2.5$ pf; $C_\pi = 78.5$ pf; $g_m = .2$ mho @ 5 MA

The small signal model for the transistor is:

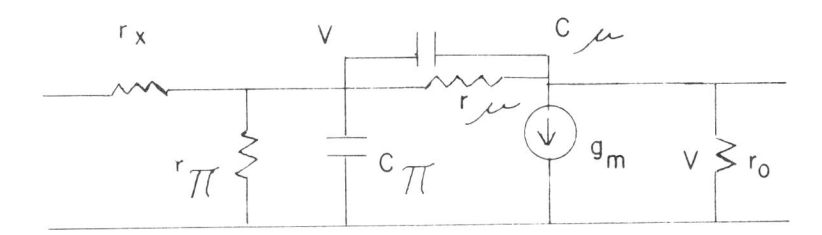

Question 2.1: Find the D.C. or quiescent operating point for each stage.

The voltage at the base, V_b, is:

$$V_b = \frac{R_2}{R_1 + R_2} V_{CC} = 2.86 \text{ v.}$$

The voltage at the emitter, V_e, is:

$$V_e = V_b - V_{be} = 2.11 \text{ v.}$$

Therefore, the collector current, I_c, is:

$$I_c = \frac{V_E}{R_4} = 5.3 \text{ mA}$$

And the collector to emitter voltage, V_{CE}, is:

$$V_{CE} = V_{CC} - I_c(R_4 + R_3) = 4.59 \text{ v.}$$

<u>Question 2.2</u>: Calculate the midband gain. To do this, first draw the small signal circuit appropriate to the mid-frequency band.

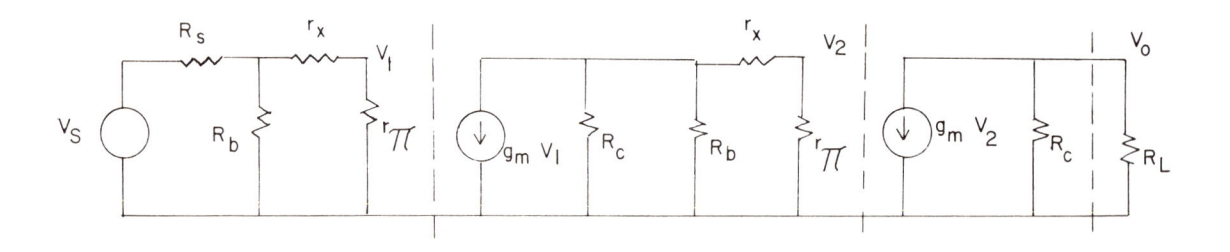

Where $R_b = R_1 || R_2 = 3.58 \text{ K}$

$R_c = r_o || R_3 = 924 \text{ }\Omega$

Then the three sections between the dotted lines are made into equivalent Thevenin circuits.

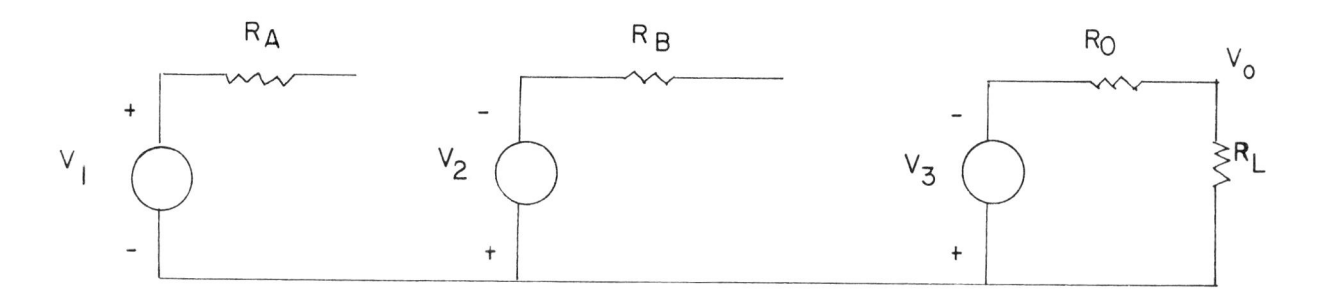

where $V_1 \simeq \dfrac{r_\pi}{r_x + r_\pi} V_s = \dfrac{400}{450} V_s$

$V_2 = -(\dfrac{r_\pi}{r_x + r_\pi})[R_c || r_b || (r_x + r)]g_m V_1$

$ = -\dfrac{400}{450}(288)\,.2V_1$

$V_o = -(R_c || R_c)g_m V_2 = -47.4(.2)V_2$

Then the midband gain V_o/V_s is:

$\dfrac{V_o}{V_s} = (\dfrac{V_o}{V_2})(\dfrac{V_2}{V_1})(\dfrac{V_1}{V_s}) = 431$

112 / ELECTRONICS

Question 2.3: Find the input impedance at mid-band.

This is the input resistance which the source generator sees:

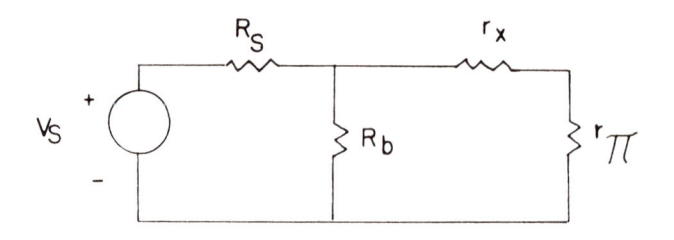

$$R_{input} = R_s + R_b || (r_x + r_\pi) = 450 \ \Omega$$

Question 2.4: Find the output impedance at mid-band.

This is just the value R_o.

$$R_o = R_c = 924 \ \Omega$$

Problem 3

This is a high impedance FET input stage amplifier.

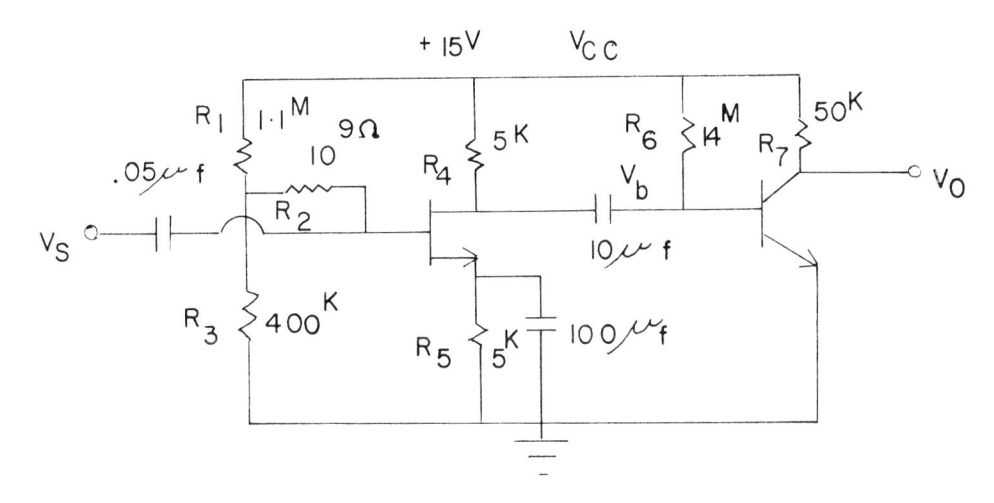

Typical specifications for an FET are shown below.

$g_m = .002$ mho

For transistor,

$r_\pi = 25$ K $\beta = 100$

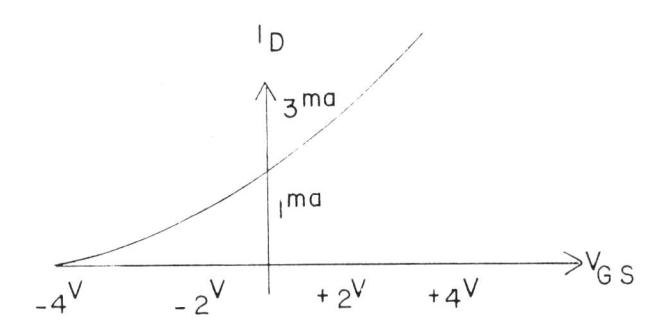

$V_{BE} = .750$ v

Question 3.1: Find the FET D.C. or quiescent operating point.

The voltage on the gate, V_G, is:

$$V_G = \frac{R_3}{R_1 + R_3} V_{CC} = 4 \text{ v.}$$

Now using the I_D vs. V_{GS} characteristic and by drawing a load line on it where R_5 is the load, the voltage from gate to source, V_{GS}, can be found.

$$V_{GS} = -1 \text{ v.}$$

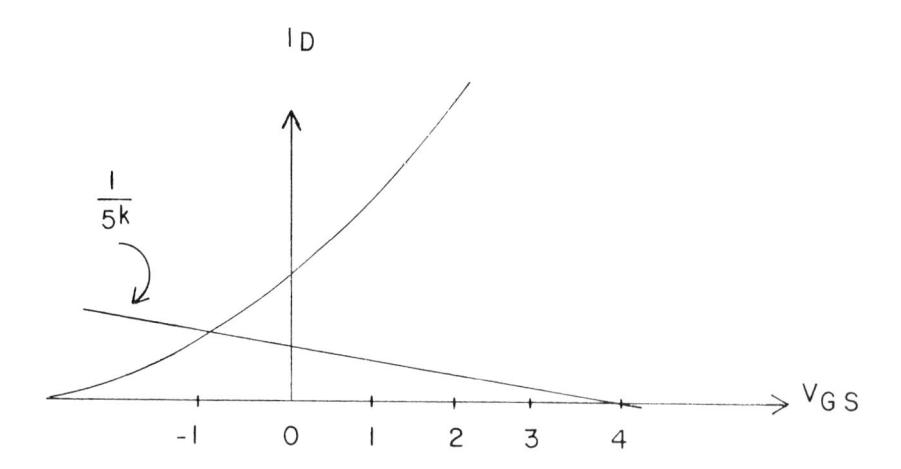

Summing voltages around the loop gives the source voltage, V_S, as

$$V_S = -V_{GS} + V_G = 5 \text{ v.}$$

Therefore, the source current, I_S, is:

$$I_S = \frac{V_S}{R_5} = 1 \text{ ma}$$

And the drain to source voltage, V_{DS}, is:

$$V_{DS} = V_{CC} - I_C(R_4 + R_5) = 5 \text{ v.}$$

Question 3.2: Find the transistor operating point.

The base current, I_B, is:

$$I_B = V_{CC} - V_{BE}/R_6 = 1 \, \mu a$$

The collector current, I_C, is:

$$I_C = \beta \, i_b = 100 \, \mu a$$

And the collector to emitter voltage, V_{CE}, is:

$$V_{CE} = V_{CC} - I_C R_7 = 10 \, v.$$

Question 3.3: Find the input impedance.

Since the FET draws no current, the input impedance is found easily as:

$$R_i = 10^9 \, \Omega$$

Question 3.4: Find the voltage gain, V_o/V_s, of the amplifier at mid-frequency band.

To do this, draw the small signal circuit.

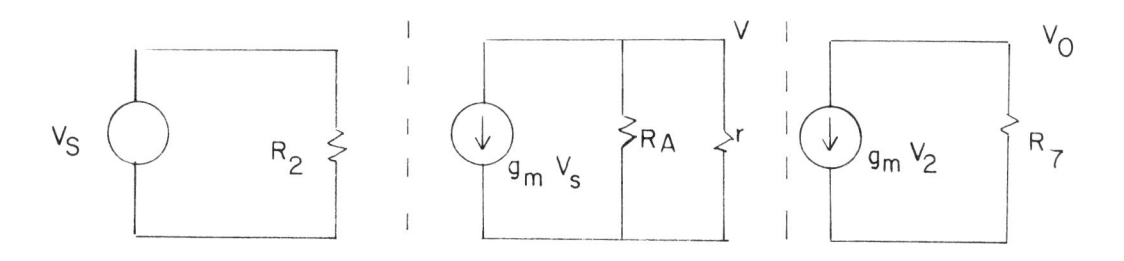

where $R_A = R_4 || R_6 = 5 \, K$

$$g_{m2} = \frac{\beta}{r_\pi} = .004 \, \text{mho}$$

Then find equivalent Thevenin circuits for the parts of the circuit between vertical dashed lines:

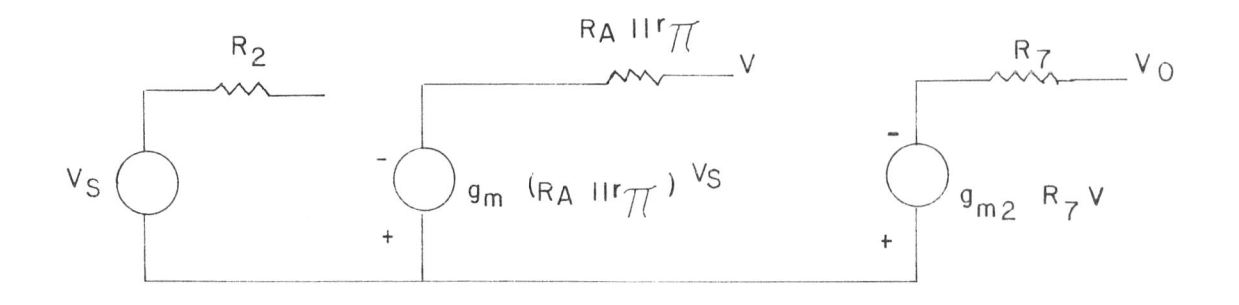

Therefore,

$$V_o = -g_{m2}R_7 V = -200\, V$$

$$V = -g_m(R_A || r_\pi)V_s = -10\, V_s$$

And the overall gain is:

$$\frac{V_o}{V_s} = 2{,}000$$

Problem 4

This is the push-pull power stage of a vacuum tube amplifier.

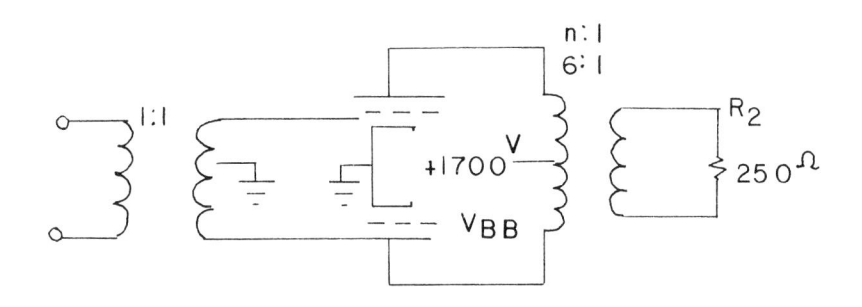

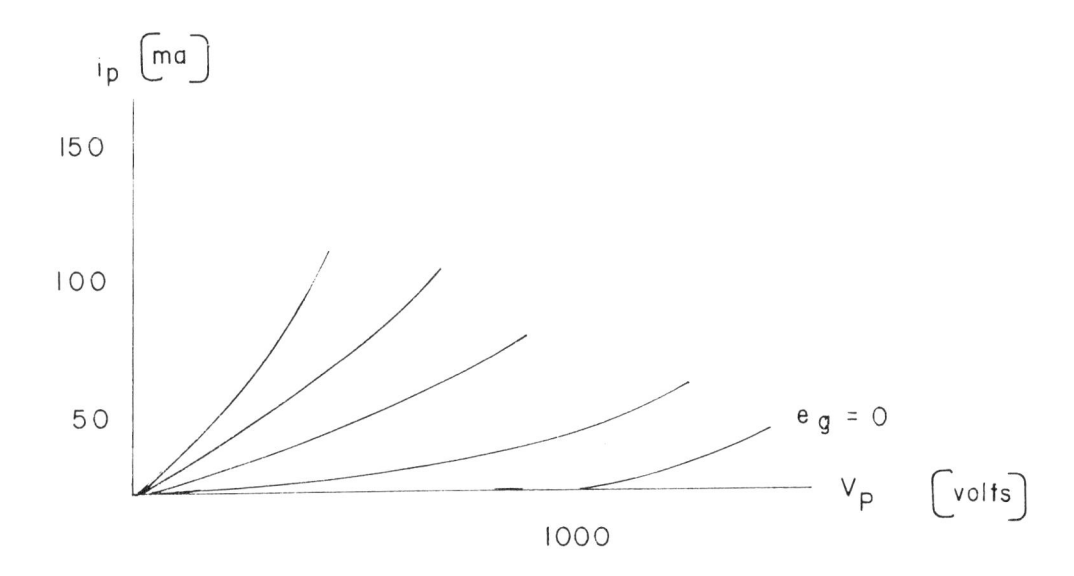

tube characteristic

Question 4.1: Find the D.C. operating point.

Since the load is transformer coupled to the stage and the impedance of the primary winding of the transformer is very small, the D.C. load line is nearly vertical. The intersection of the load line with the $e_g = 0$ characteristic gives the D.C. operating point as:

$$v_p \approx 1,200 \text{ v.} \qquad i_p \approx 0$$

118 / ELECTRONICS

Question 4.2: **Draw** the AC load line on the v_p vs. i_p characteristics.

First find the load reflected to the primary of the transformer, R'_L, as:

$$R'_L = n^2 R_L = 9 \text{ K}$$

Then find the current that load draws at no voltage, i_{pmax}, as:

$$i_{pmax} = \frac{V_{BB}}{R'_L} = 133 \text{ ma}$$

Then draw the load line.

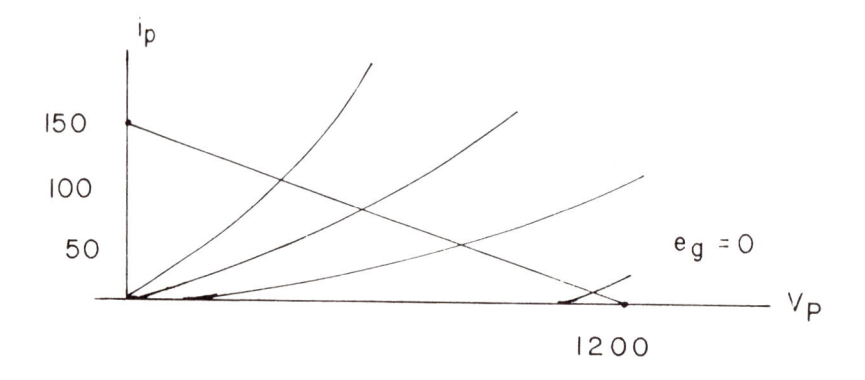

Question 4.3: Assuming a sinusoidal drive signal into the stage, what is the maximum average power into the load, P_L?

Draw the wave forms of load current and voltage.

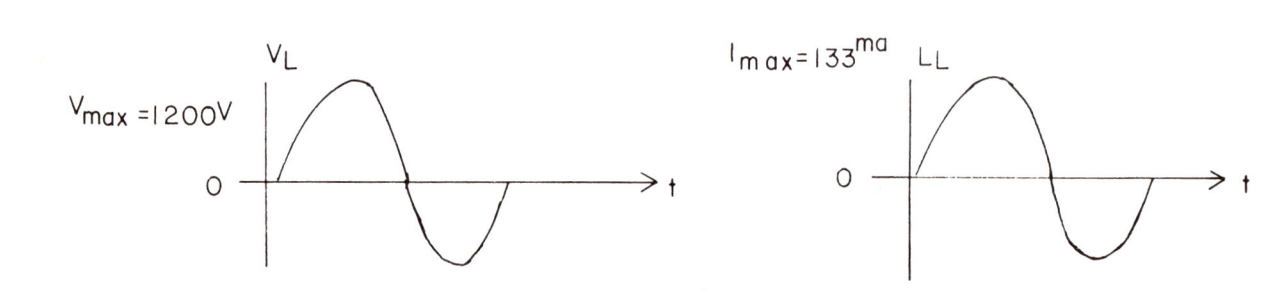

ELECTRONICS / 119

Since the instantaneous power, P_i, is:

$$P_i = v_L i_L$$

Then the average load power is:

$$P_L = V_{max} I_{max} < \sin^2 wt >$$

where $< >$ means time average of the inside function.

And since

$$< \sin^2 wt > = \frac{1}{2}$$

then

$$P_L = \frac{1}{2} V_{max} I_{max} = 79.8 \text{ watts}$$

Question 4.4: Find the average power drawn from the D.C. supply, P_s.
Draw the waveforms of supply voltage and current.

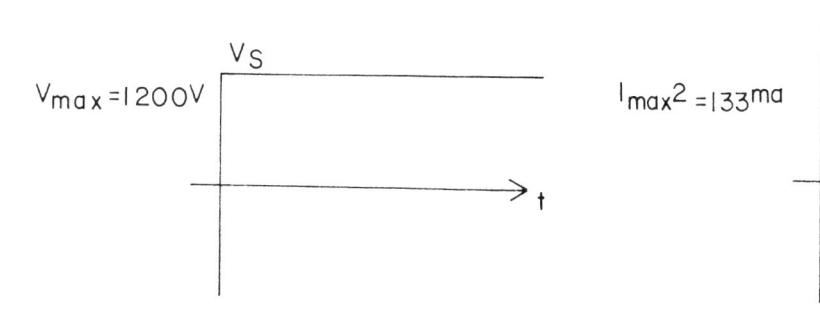

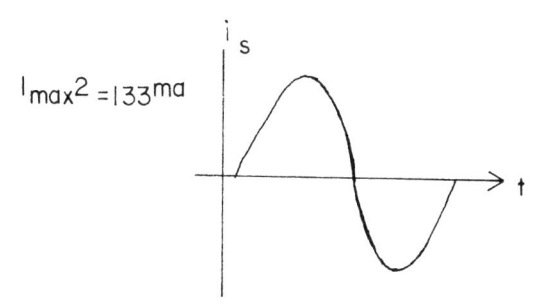

$V_{max} = 1200V$

$I_{max2} = 133 \text{ ma}$

120 / ELECTRONICS

Then the instantaneous power, P_{is} is:

$$P_{is} = V_s i_s$$

and the average power from the supply, P_{AVS}, is:

$$P_{AVS} = V_{max} I_{max} <\sin wt>$$

To find $<\sin wt>$, perform the following averaging integration:

$$<\sin wt> = \frac{1}{\pi} \int_0^\pi \sin\theta \, d\theta = \frac{1}{\pi} \left| \cos\theta \right|_\pi^0 = \frac{2}{\pi}$$

Therefore

$$P_{AVS} = \frac{2}{\pi} V_{max} I_{max} = 102 \text{ watts}$$

Question 4.5: Find the efficiency of power transfer for the stage defined as

$$n \triangleq \frac{P_{AV}}{P_{AVS}} .$$

$$n = \frac{P_{AV}}{P_{AVS}} \times 100\% = 78\%$$

Problem 5

This is a three-stage direct coupled transistor amplifier.

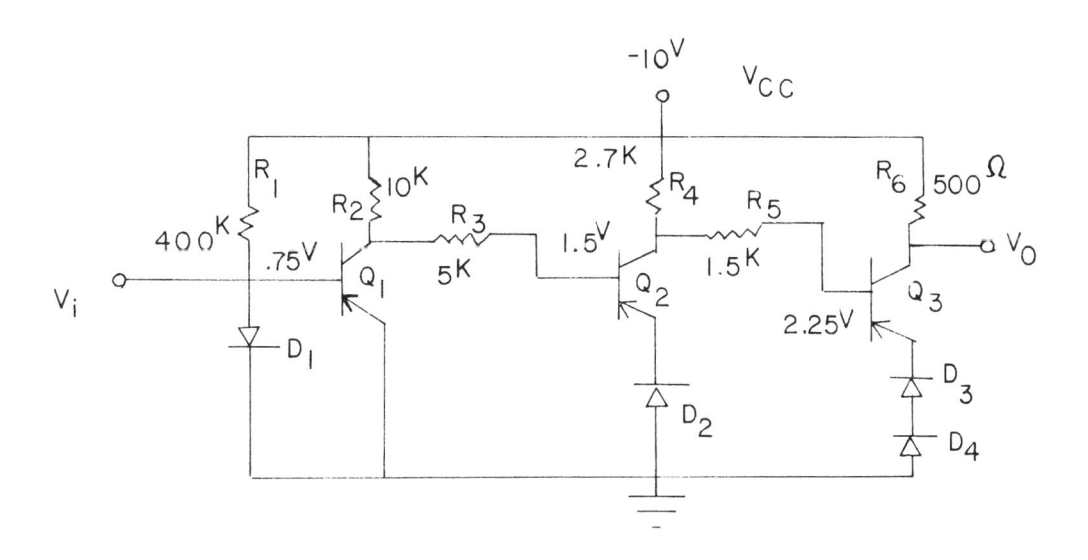

forward diode drop, V_δ = 750 mv; V_{EB} = 750 mv; r_π = 300 Ω; β = 32; I_{CO} = 1µa; α = .97.

Question 5.1: Find the leakage current, I_D, which must be present through D1 to compensate for the I_{co} through Q1.

The model of the transistor is the following:

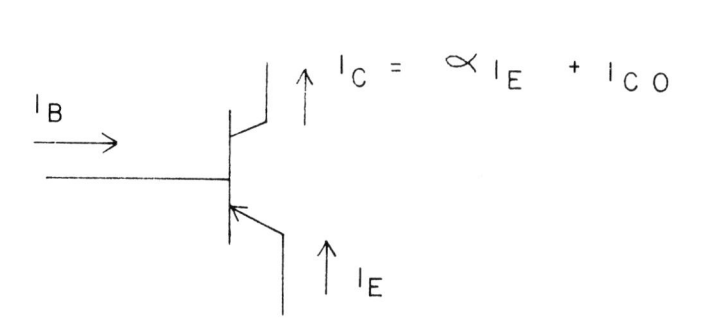

$$I_C = \alpha I_E + I_{CO}$$

To get I_c in terms of I_B, write a node equation.

$$I_E = I_c - I_B$$

Substitute this into expression for I_c.

$$I_c = \alpha(I_c - I_B) + I_{co}$$

or

$$I_c = -\frac{\alpha I_B}{1-\alpha} + \frac{I_{co}}{1-\alpha}$$

Now write down what I_B is.

$$I_B = \frac{V_{CC} - V_{BE}}{R_1} + I_D = I + I_D$$

Substitute into expression for I_c.

$$I_c = -\frac{\alpha I}{1-\alpha} + \frac{I_{co} - \alpha I_D}{1+\alpha}$$

To compensate for transistor I_{co},

$$I_{co} = \alpha I_D \qquad \text{or} \qquad I_D = \frac{I_{co}}{\alpha}$$

$$I_D = 1.03 \, \mu a$$

Question 5.2: Find the D.C. operating point of the first stage.

$$I_{B1} = \frac{V_{CC} - V_{BE}}{R_1} = 23.1 \, \mu a$$

$$I_{c1} = \beta I_{B1} = 740 \, \mu a$$

Then find the current through R_2 to get V_{cE1} by:

$$(I_{B2} + I_{c1})R_2 + I_{B2}R_3 = V_{CC} - 2V_{BE}$$

or

$$I_{B2} = 73.5 \, \mu a$$

The current through R_2 is:

$$I_{R2} = I_{c1} + I_{B2} = 813.5 \ \mu a$$

Therefore

$$V_{cE1} = V_{CC} - I_{R2}R_2 = 1.12 \ v.$$

Question 5.3: Find the D.C. operating point for the second stage.

$$I_{c2} = \beta I_{B2} = 2.35 \ ma$$

Then find the current through R_4 to get V_{cE2} by:

$$(I_{B3} + I_{c2})R_4 + I_{B3}R_5 = V_{CC} - 3V_{BE}$$

or

$$I_{B3} = 333 \ \mu a$$

The current through R_4 is:

$$I_{R4} = I_{c2} + I_{B3} = 2.68 \ ma$$

Therefore

$$V_{cE2} = V_{CC} - I_{R4}R_4 - V_{\gamma}$$

$$V_{cE2} = 2.35 \ v.$$

Question 5.4: Find the D.C. operating point of the third stage.

$$I_{c3} = \beta I_{B3} = 10.6 \text{ ma}$$

$$V_{cE3} = V_{CC} - I_{c3}R_6 - 2V\gamma$$

$$V_{cE3} = 3.2 \text{ v.}$$

Question 5.5: Find the small signal gain V_o/V_s.

Draw the small signal circuit model.

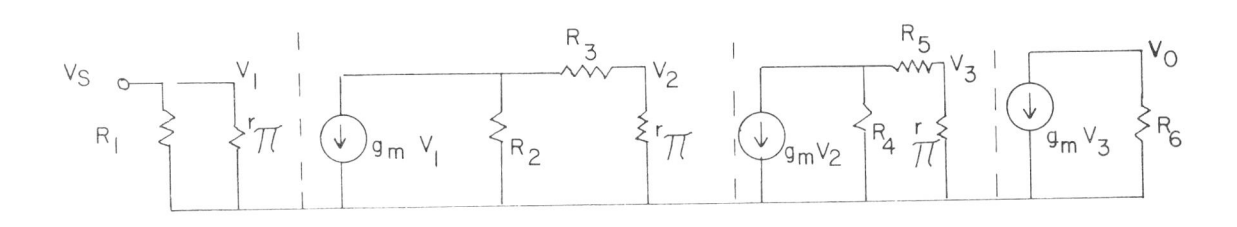

Find the voltage ratios, V_1/V_s; V_2/V_1; V_3/V_2; V_o/V_3 by using Thevenin's theorem.

$$V_1/V_s = 1$$

$$V_2/V_1 = -\frac{R_2 g_m r_\pi}{R_2 + R_3 + r_\pi} = -20.2$$

$$V_3/V_2 = -\frac{R_4 g_m r_\pi}{R_4 + R_5 + r_\pi} = -17.3$$

$$V_o/V_3 = -g_m R_6 = -20$$

The overall gain then is:

$$V_o/V_s = -7,000$$

References

Schaum's Outline Series   Electronic Circuits, Chapters 2, 5, 6, and 7.

Quantum Publishers, Incorporated, Series on Electronic Circuits, Chapters 2, 3, 4, 6, and 8.

Chapter 9

ILLUMINATION

Problem 1

Purpose and Background: The purpose of this problem is to illustrate the use of the Inverse Square Law in finding the illumination at a point on a surface produced by a point source of light. It is assumed that the light travels directly to the receiving surface with no reflections off other surfaces.

Reference: IES Lighting Handbook, 5th edition, section 4.

Statement of Problem: A long straight street is lighted by mercury street lighting units on poles 20 feet high and 100 feet apart. Each lighting unit has an intensity of 2,000 candelas when viewed from any point along the street. Find the following:

a) illumination on the street at a point midway between two poles.

b) illumination on a small vertical sign five feet above the street and midway between two poles.

The pertinent equation to use is

$$E = \frac{I}{d^2} \cos\theta \qquad (1)$$

where E is the illumination in footcandles (lumens per square foot), I is the intensity of the source in candelas, d is the distance in feet from the source to the point P at which the illumination is desired, and θ is the angle of incidence, the angle between a line drawn perpendicular to the surface from point P and a line drawn from the source to the point P.

128 / ILLUMINATION

Solution:

a)

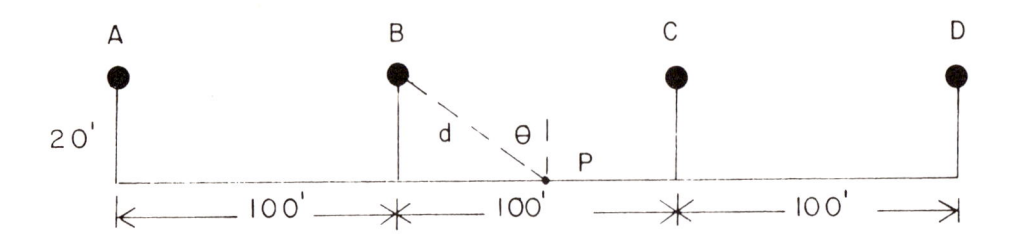

```
        A              B              C              D
        ●              ●              ●              ●
    20'                 \d   θ|
                         \   |
                          \  | P
        |——— 100' ———*——— 100' ———*——— 100' ———|
```

Figure 1

Referring to Figure 1, the horizontal illumination at point P is due predominantly to lamps B and C, but we should also check the contribution of lamps A and D. For lamps B and C,

$$d = \sqrt{50^2 + 20^2} = 53.85 \text{ feet}$$

$$\cos \theta = 20/53.85 = .371$$

Thus,

$$E = 2 \times \frac{2,000}{2,900} \times .371 = .51 \text{ ft-c}$$

For lamps A and D,

$$d = \sqrt{150^2 + 20^2} = 151.3 \text{ feet}$$

$$\cos \theta = 20/151.3 = .132$$

and

$$E = 2 \times \frac{2,000}{2,290} \times .132 = .02 \text{ ft-c}$$

The total illumination at point P is .53 ft-c. As can be seen, lamps A and D contribute negligibly to the total illumination. We should expect this, not only because they are further away but because they have a much larger angle of incidence with the surface at point P.

b)

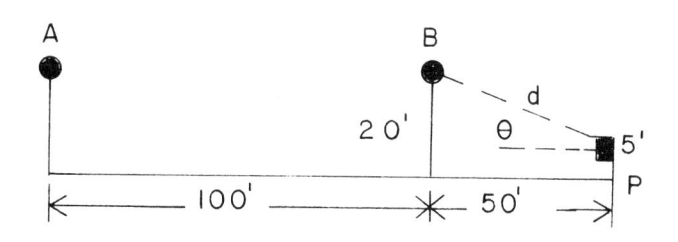

Figure 2

The situation is shown in Figure 2. For lamp B, we have

$$d = \sqrt{50^2 + 15^2} = 52.20 \text{ feet}$$

$$\cos \theta = 50/52.20 = .958$$

Then, using Equation (1),

$$E = \frac{2,000}{2,725} \times .958 = .70 \text{ ft-c}$$

For lamp A, the calculations are

$$d = \sqrt{150^2 + 15^2} = 150.75 \text{ feet}$$

$$\cos \theta = 150/150.75 = .995$$

and Equation (1) yields

$$E = \frac{2,000}{22,725} \times .995 = .09 \text{ ft-c}$$

The total illumination on the sign is .79 ft-c.

130 / ILLUMINATION

Problem 2

Purpose and Background: The purpose of this problem is to help the reader become familiar with the various terms and units used in illumination engineering. Specifically, the problem will involve intensity in candelas, luminous flux in lumens, illumination in footcandles (lumens per square foot), and luminance in footlamberts (lumens per square foot) and in candelas per square inch. The footcandle and footlambert have the same dimensions. The footcandle refers to lumens per square foot <u>hitting</u> a surface; the footlambert to lumens per square foot <u>leaving</u> a surface. Also, over the years, two units of luminance have emerged -- the footlamber and the candela per square inch. We have defined the former. The latter is the luminous intensity in candelas of a source or surface in a particular direction per unit of <u>projected</u> area of the source or surface in that direction. Projected area is illustrated in Figure 3, which shows an end view of a flat square surface d feet on a side and inclined at an angle θ with the vertical.

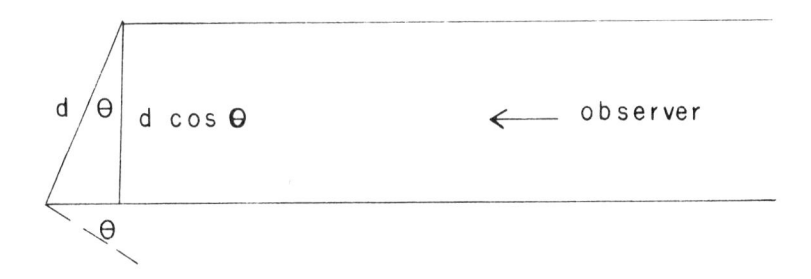

Figure 3

If an observer stands some distance from the surface, he sees a surface whose area appears to be $d^2 \cos\theta$ square feet.

The two units of luminance can be related for a surface that is perfectly diffusing. For such a surface, luminance is constant with viewing angle and the luminance in footlamberts is obtained by multiplying the luminance in candelas per square inch by 144π.

Reference: <u>IES Lighting Handbook</u>, 5th edition, sections 1, 4.

Statement of Problem: A translucent spherical glass globe is 15 inches in diameter and has a luminance of 3 candelas per square inch. It houses an incandescent lamp at its center with an intensity of 700 candelas, uniform in all directions. Assuming the globe to be perfectly diffusing, find the following:

a) luminous flux emitted by the lamp.

b) illumination on the inside of the globe.

c) luminance of the globe in footlamberts.

d) transmission factor of the globe.

Solution:

a) A one candela source emits 4π lumens of luminous flux. Therefore, the incandescent lamp emits

$$700 \times 4\pi = 2,800\pi \text{ lumens.}$$

b) The surface area of a sphere is $4\pi r^2$. The globe radius is .625 feet and its surface area is 1.56π square feet. Thus, the illumination on the inside of the sphere is

$$E = \frac{2,800\pi}{1.56\pi} = 1,800 \text{ ft-c}$$

c) The globe has a luminance of 3.00 candelas per square inch. Since the globe is perfectly diffusing, its luminance in footlamberts is

$$144\pi \times 3 = 1,360 \text{ footlamberts}$$

d) There are 1,360 lumens per square foot leaving the outside of the globe and 1,800 lumens per square foot hitting the inside of the globe. The ratio of these is the globe's transmission factor.

$$tf = \frac{1,360}{1,800} = .76 = 76\%$$

We can also obtain tf from the luminance in candelas per square inch. The projected area of the globe is $\pi r^2 = 176.7$ square inches. Thus, the intensity of the outside of the globe is

$$176.7 \times 3 = 530 \text{ candelas.}$$

The inside intensity is 700 candelas, giving

$$tf = \frac{530}{700} = .76$$

132 / ILLUMINATION

Problem 3

Purpose and Background: The purpose of this problem is to consider a situation where intensity varies with angle. This is the usual situation particularly in floodlighting and spotlighting applications.

Reference: *IES Lighting Handbook*, 5th edition, sections 4, 8.

Statement of Problem:

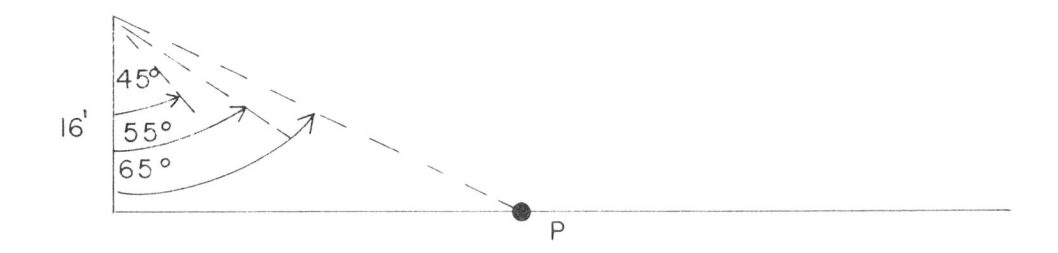

Angle	Candelas
0	16,000
10	14,000
20	10,000
30	2,000
40	500
50	0

Figure 4

A shuffleboard runway is to be lighted by three PAR lamps on a 16-foot high pole aimed as shown in Figure 4. Find the horizontal illumination at point P, on the axis of the lamp aimed at 65°.

Solution: The Inverse Square Law (Equation 1) will be used but we must be careful to insert the correct intensity values. For the lamp aimed at P, we use 16,000 candelas. For the other two lamps, we use 14,000 candelas (10° from the beam center) and 10,000 candelas (20° from the beam center).

The solution proceeds as follows:

$$\cos 65° = .423 = 16/d$$

$$d = 37.8 \text{ feet}$$

$$E = \frac{16{,}000}{37.8^2} \times .423 + \frac{14{,}000}{37.8^2} \times .423 + \frac{10{,}000}{37.8^2} \times .423$$

$$= \frac{40{,}000}{37.8^2} \times .423 = 11.8 \text{ ft-c}$$

Problem 4

Purpose and Background: When dealing with interior spaces, it is no longer accurate to use the Inverse Square Law to compute illumination levels. Multiple reflections from ceilings, walls, and floors cause the actual illumination to be higher than the Inverse Square Law would predict. There are techniques for including these reflections in the calculation of illumination at a point in an interior space. However, by far the preferred approach has been to obtain the average maintained footcandle level throughout a room, rather than the illumination at several points within a room. The standard procedure for obtaining the former is the Lumen-Zonal Cavity Method of interior lighting design. It is the purpose of this problem to illustrate that procedure. The Lumen-Zonal Cavity Method is based on the premise that one can obtain a "coefficient of utilization" (CU) which is the ratio of the total lumens reaching the work plane (an imaginary plane parallel to the floor and generally 30 inches above it) to the total lumens from the lamps present. Thus, CU takes into account absorptions by the luminaires (lighting fixtures) and room surfaces.

In addition to CU, if we desire a certain maintained footcandle level rather than merely an initial footcandle level, we need to include depreciation. This we do with a light loss factor (LLF) which includes room surface dirt accumulation (RSDD), lamp burnouts which are not immediately replaced (LBO), luminaire dirt accumulation (LDD), and lamp lumen depreciation as the lamp ages (LLD).

In equation form, we can include CU and LLF and write

$$E = \frac{\phi \times CU \times LLF}{A} \qquad (2)$$

134 / ILLUMINATION

where E is the average maintained illumination level on the work plane, ϕ is the total initial lamp lumens, and A is the area of the work plane. If we exclude LLF, then we obtain the average <u>initial</u> illumination level on the work plane.

<u>Reference</u>: <u>IES Lighting Handbook</u>, 5th edition, sections 8 and 9.

<u>Statement of Problem</u>: A school room, 20 feet by 45 feet with a 10-foot ceiling, is to be lighted by 40-watt (4 feet long) standard cool white fluorescent rapid start lamps using Unit No. 31 (IES Handbook, page 9-24). Assume two lamps per fixture and that the fixtures are suspended 1.5 feet from the ceiling. Room reflectance values are ceiling, 80%; walls, 50%; floor, 30%. An average maintained illumination level of 50 footcandles is desired on a work plane 30 inches above the floor (see <u>IES Handbook</u>, pages 9-81 to 9-95 for recommended illumination levels).

<u>Solution</u>:

Step 1 - Cavity Ratios
The Lumen-Zonal Cavity Method breaks a room up into three spaces (cavities), a ceiling cavity, a room cavity, and a floor cavity (see Figure 5). It is necessary to obtain a cavity ratio for each of these spaces. This may be done using a table in the <u>IES Handbook</u> (page 9-9) or from the following equation:

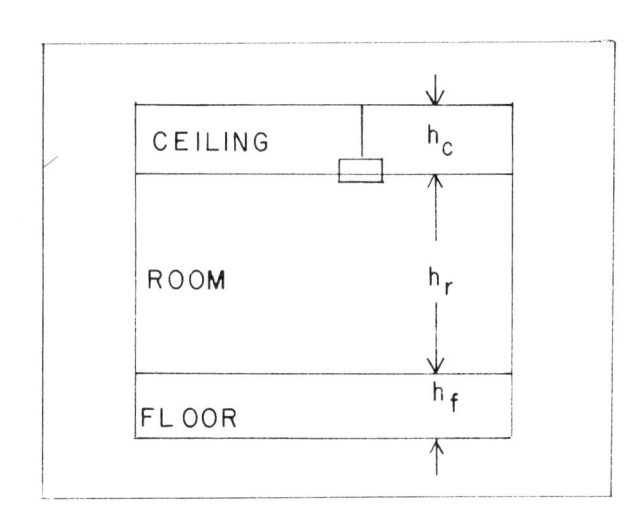

Figure 5

$$CR = \frac{5h(\ell + w)}{\ell \times w} = \frac{2.5hP}{A} \qquad (3)$$

In this equation, h is cavity height and is h_c for the ceiling cavity, h_r for the room cavity, and h_f for the floor cavity, ℓ and w are the room length and width, P is room perimeter, and A is the work plane area. The first form of the equation is used for rectangular rooms; the second form for rooms of irregular shape.

For the particular problem at hand,

$$RCR = \frac{5 \times 6 \times 65}{900} = 2.17$$

$$CCR = \frac{5 \times 1.5 \times 65}{900} = .54$$

$$FCR = \frac{5 \times 2.5 \times 65}{900} = .91$$

Step 2 - Effective Cavity Reflectances
It is now necessary to obtain the effective cavity reflectance of the ceiling and floor cavities. For the ceiling cavity, what we desire is the effective reflectance of the imaginary horizontal plane drawn through the fixtures; for the floor cavity, we need the effective reflectance of the work plane. Then we can treat these cavity reflectances, along with the wall reflectance, as the reflectances of a "new" room -- the room cavity of height h_r.

The two effective reflectances are obtained from the table on pages 9-10 and 9-11 in the IES Handbook. Their values are

$$\rho_{cc} = .72 \qquad \rho_{fc} = .27$$

Step 3 - Coefficient of Utilization
Coefficients of utilization are given on pages 9-12 through 9-30 of the IES Handbook for the 50 fixture types listed. Entering this table for Unit No. 31 with a RCR of 2.17, a ρ_{cc} of .72, and a ρ_w of .50 yields, after interpolation, a value of CU = .59. This value of CU is based on a 20% floor cavity reflectance; our value is $\rho_{fc} = .27$. Correction factors for other floor cavity reflectances are given on page 9-32 of the IES Handbook. The correction factor for a 30% floor cavity reflectance, again after interpolation, is 1.057. Taking .7 of the increment yields a correction factor of 1.040 and a final CU value of 1.04 x .59 = .61.

Step 4 - Light Loss Factor
As mentioned earlier, this factor is made up of four parts.

RSDD -- We will assume a clean room which is cleaned once annually. Then from Figure 9-5 (page 9-4) in the IES Handbook and noting that Unit No. 31 is a semi-direct luminaire, we obtain RSDD = .95.

LBO -- We will assume that burned out lamps are replaced promptly. Thus, LBO = 1.00.

LDD -- Included with the data for Fixture No. 31 is a maintenance category V. Again assuming a clean room which is cleaned once annually, we can use Figure 9-7 (page 9-6) in the IES Handbook to obtain an LDD factor of .88.

LLD -- Lamp data is presented in an extensive table at the close of section 8 in the IES Handbook. Our lamp appears on page 8-94 and is a 40T12 (40 watts, tubular, and 12/8 = 1 1/2 inches in diameter). Its initial lumen output is 3,200 and its LLD factor is listed as .84 which is the fraction of initial lumen output remaining at 70% life (.7 x 18,000 = 12,600 hours), based on 3 burning hours per start.

Multiplying these four numbers together gives

LLF = .95 x 1.00 x .88 x .84 = .70

Step 5 - Number of Lamps (N)
We can now use Equation 2 to find the total initial lumens required.

$$\phi = \frac{50 \times 20 \times 45}{.61 \times .70} = 105,400 \text{ lumens}$$

With each lamp providing 3,200 initial lumens, we require

$$N = \frac{105,400}{3,200} = 33 \text{ lamps}$$

or 17 two-lamp fixtures.

Step 6 - Lighting Layout
The first step in making the lighting layout is to determining the maximum spacings allowed between fixtures. Included in the data for Fixture No. 31 is a maximum spacing to mounting height above work plane ratio of 1.2. If we locate the fixtures no further apart than 1.2 x 6 = 7.2 feet, then the maximum and minimum illumination values on the work plane will be within ±1/6 (±17%) of the average value.

We must also concern ourselves with the distance of a fixture from the walls. To prevent drop off of illumination near the walls, where desks and tables may be located, a fixture should be no more than 2 feet from its end wall and no more than 2 1/2 feet from its side wall.

A possible layout for the problem at hand, taking into account the number of lamps needed and the spacing criteria, is shown in Figure 6.

```
                            45'
         ┌─────────────────────────────────────────────┐
         │  3'____3'__2½'__3'_____3'_____3'_____3'____3'│
         │                                             │
         │         7½'                                 │
         │                                             │
         │      ___    ___   ___    ___   ___    ___   │
    20'  │                                             │
         │         7½'                                 │
         │                                             │
         │      ___    ___   ___    ___   ___    ___   │
         │         2½'                                 │
         └─────────────────────────────────────────────┘
```

Figure 6

This layout provides 36/33 x 50 = 55 footcandles of average maintained illumination, and it is very close to satisfying the spacing criteria for reasonably uniform illumination. However, it does require that the fixture be mounted individually rather than in continuous rows which could increase installation cost and make fixture alignment difficult.

Probably a better solution would be to mount the fixtures on the ceiling. This changes the CU slightly to .62, requiring now 104,000 lumens, 32 lamps, and 16 fixtures. The maximum spacing is now 1.2 x 7.5 = 9 feet. By giving up a bit on uniformity, we could now go to two continuous rows, spaced, say, 12 feet apart, 4 feet from each side wall, and 6 1/2 feet from each end wall. Or, if one side wall were largely windows, we could space the first row of units, say, 7 feet from the windows and allow 10 feet between rows and 3 feet from the inside row to its side wall.

Step 7 - Energy Consumption

Assuming we are using the layout in Figure 6, we require 18 two-lamp fluorescent units. To ascertain the power requirement, we use Figure 8-115 (page 8-100) in the <u>IES Handbook.</u> This table gives the wattage consumptions for the various types of fluorescent lamps and their ballasts. For the 40 watt rapid start lamp, each 2-lamp unit, including ballast, consumes 92 watts. For the total installation, the power required is 18 x 92 = 1,656 watts, giving 1,656/900 = 1.8 watts per square foot and 1,656/55 = 30 watts per footcandle.

138 / ILLUMINATION

These are the two customary measures of the efficiency of a lighting design and should be computed for each trial design.

Problem 5

Purpose and Background: Very often when dealing with line and area sources, the Inverse Square Law cannot be used directly because the maximum source dimension is appreciable with respect to the distance from the source to the receiver surface at which the illumination is to be obtained. In other words, the source does not appear as a point source.

In this problem we will discuss how to obtain the illumination from a non-point source. What we do is break the source up into differential source elements that are small enough to be considered as point sources, use the Inverse Square Law to express the illumination at the chosen point produced by the differential element, and integrate over the source to obtain the total illumination

The general situation is shown in Figure 7. We assume that

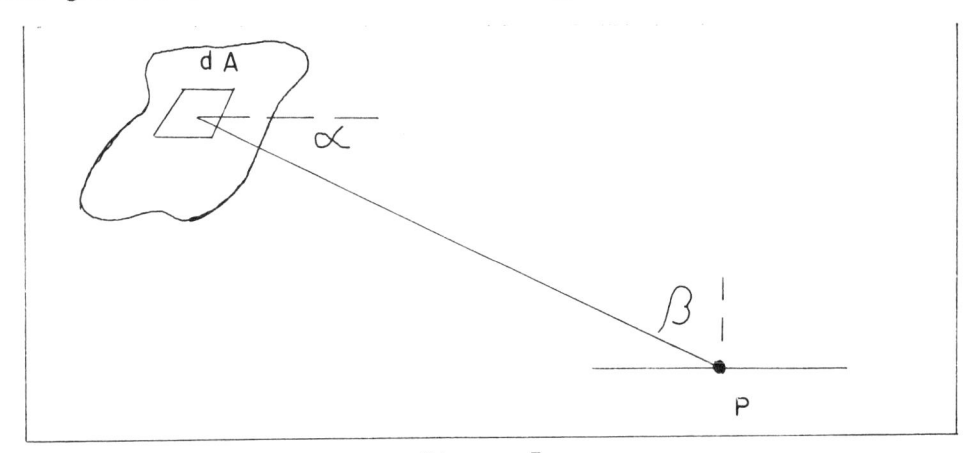

Figure 7

the line from source to receiver makes an angle α with the perpendicular to the source element and an angle β with respect to the perpendicular to the receiver surfact at point P. The intensity of the source element in the direction of P is given by

$$dI = L \cos \alpha \, dA \qquad (4)$$

where L is the luminance of the source element in candelas per unit area and cos α dA is the projected area of the source element as viewed from P.

Now using Equation 1, we have

$$dE = \frac{L \cos \alpha \cos \beta \, dA}{d^2} \qquad (5)$$

for the illumination at P produced by dA. Integrating yields

$$E = \int_s \frac{L \cos \alpha \cos \beta \, dA}{d^2} \qquad (6)$$

References: IES Lighting Handbook, 5th edition, section 9.
"Illumination Engineering," W.B. Boast, McGraw-Hill, chapter 7.

Statement of Problem: A circular disk source one foot in radius lights a parallel plane 4 feet below it. Find the illumination at a point in the plane directly below the source's center if the source luminance is 4 candelas per square inch, uniform over the entire source surface.

Solution:

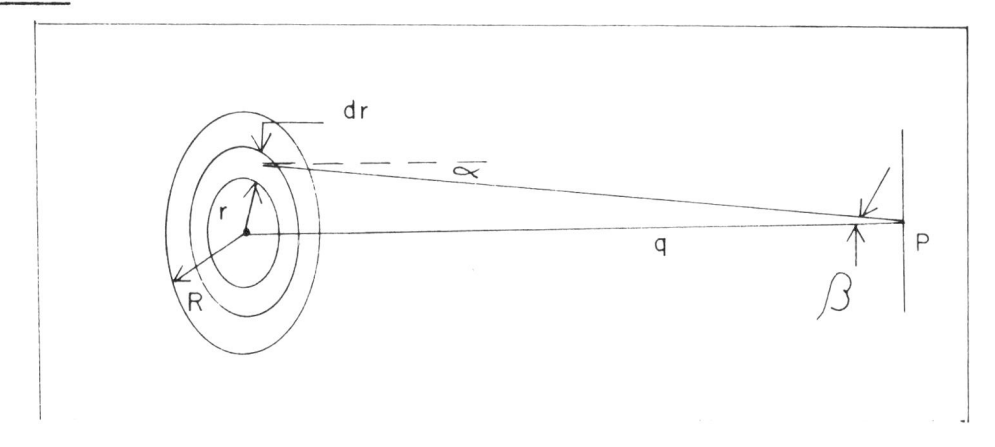

Figure 8

The situation is shown in Figure 8.

From Equation 5 we can write

$$dE_p = \frac{L \, q \, q \, 2\pi r \, dr}{\sqrt{q^2 + r^2} \sqrt{q^2 + r^2} \, (q^2 + r^2)} = \pi q^2 L \frac{2r \, dr}{(q^2 + r^2)^2}$$

140 / ILLUMINATION

Integrating from 0 to R yields

$$E_p = -\pi q L \left. \frac{1}{q^2 + r^2} \right|_0^R = \frac{\pi L R^2}{q^2 + R^2}$$

Inserting values gives

$$E_p = \frac{\pi \times 4 \times 144 \times 1}{16 + 1} = 106 \text{ ft-c}$$

It should be noted that if $q \gg R$, we obtain

$$E_p \approx \frac{\pi L R^2}{q^2} = \frac{LA}{q^2} = \frac{I}{q^2} = 113 \text{ ft-c}$$

In other words, we return to what Equation 1 would give us if the Inverse Square Law held for the entire source. It is interesting to prepare a chart comparing actual illumination with Inverse Square Law illumination for various values of q.

q (ft.)	E_{actual} (ft-c)	E_{ISL} (ft-c)
.5	1,448	7,238
1	905	1,810
2	362	452
4	106	113
6	49	50
8	28	28

As a rough rule of thumb, if the distance from source to receiver is 5 times the maximum source dimension, or greater, there will be no appreciable error through use of the Inverse Square Law in Equation 1.

Problem 6

Purpose and Background: This is a problem in lighting economics. There are two major costs in a lighting installation -- the initial investment cost and the annual owning and operating cost. It is the purpose of this problem to calculate each of these for a typical lighting installation.

ILLUMINATION / 141

Reference: <u>IES Lighting Handbook</u>, 5th edition, section 15

<u>Statement of Problem</u>: a) For the following lighting installation, compute the initial investment cost and the overall owning and operating cost.

> Lamp - 40 watt rapid start fluorescent
> Lamps per Luminaire - 2
> Number of Luminaires - 100
> Energy rate - $0.05 (5¢) per KWH
> Burning hours per year - 2,000
> Luminaire cost (each) - $51.00
> Installation cost per luminaire - $10.00
> Lamp cost (list less 35%) - $0.84 (84¢)
> Labor cost of lamp replacement (each) - $1.50
> Cleaning cost per luminaire - $2.00
> Cleanings per year - 1
> Annual fixed charges - 15% of initial cost

b) Assume the average illumination level is 100 footcandles and the room dimensions are 60 feet by 60 feet. Find the annual costs per footcandle and per square foot.

<u>Solution</u>:

a) Initial Investment Cost

Luminaire cost - $51.00 x 100	=	$ 5,100.00
Installation cost - $10.00 x 100	=	1,000.00
Lamp cost - $0.84 x 200	=	168.00
		$ 6,268.00

Annual Fixed Charges - 15% of initial cost, less lamps

.15 x $6,100.00 = $915.00

(Since lamps are "consumed," it is not correct to include them in the annual fixed charges on the capital investment.)

Annual Maintenance Cost

Lamp Life (from <u>IES Handbook,</u> Pages 8-94) - 18,000 hours

Annual Lamp replacement cost

$$\frac{200 \times 2,000}{18,000} = 22.2 \text{ lamps}$$

22.2 x $0.84 = $18.65

142 / ILLUMINATION

>Annual labor cost of lamp replacement
>
>22.2 x $1.50 = $33.30
>
>Annual cleaning cost
>
>100 x $2.00 x 1 = $200.00
>
>Total annual maintenance cost = $251.95

Annual Energy Cost

>Watts per luminaire - 92 watts (see page 8-101 or IES Lighting Handbook)
>
>Total watts = 9,200
>
>Energy cost annually
>
>$$\frac{9,200 \times \$0.05 \times 2,000}{1,000} = \$920.00$$
>
>Annual operating cost = $1,171.95
>Annual owning cost = 915.00
>Annual owning and operating cost = $2,086.95

b) $\frac{\$2,086.95}{100} = \$20.87$ per ft-c

$\frac{\$2,086.95}{3,600} = \$0.58$ (58¢) per square foot

Problem 7

Purpose and Background: It is the purpose of this problem to assist the reader in updating his understanding of the three major types of light sources -- incandescent, fluorescent, and high intensity discharge. This will be done through a series of qualitative questions about incandescent, fluorescent, mercury, metal halide, and high pressure sodium lamps.

Reference: IES Lighting Handbook, 5th edition, section 8

Questions and Answers:

1. Q. Distinguish between a flood-lamp and a spot lamp; between a reflector (R) lamp and a projector (PAR) lamp. (Handbook - pages 8-11 and 8-12)

ILLUMINATION / 143

1. A. A floodlamp has a wider beam than a spot lamp, approximately $\pm 30°$ for a flood and $\pm 15°$ for a spot out to 10% of maximum beam intensity. Also, for a given wattage, a spot has a much higher end-on intensity than a flood.

    An R lamp is a one-piece blown bulb to which a reflectorized coating is applied. A PAR (parabolic aluminized reflector) lamp has a two-piece construction -- a reflector portion and a lens cover sealed to it during manufacture. Of the two, the PAR lamp is much more mechanically durable.

2. Q. Explain the difference between a tungsten-halogen lamp and a conventional incandescent lamp. (Handbook - page 8-6)

    A. Tungsten-halogen lamps utilize a halogen element, such as iodine, inside them. Tungsten particles boiled off the filament combine with the iodine to form tungsten-iodide rather than depositing on the bulb wall. As the tungsten-iodide particles circulate back toward the filament, temperatures are such that the tungsten and iodine separate with the former redepositing on the filament. The result is a considerable improvement in lamp life and a noticeable reduction in lumen depreciation.

3. Q. Distinguish between preheat, rapid start, and instant start fluorescent lamps. (Handbook - pages 8-28 and 8-29)

    A. Preheat lamps were the first fluorescent lamps introduced. They require preheating of the cathode at each end of the lamp before the arc is struck. This is accomplished either manually, with a push-button switch, or automatically with a small neon glow tube starter.

    Instant start lamps came next, and were introduced by the lamp companies to rid the user of the annoying delay in starting preheat lamps. Instant start lamps replace preheating by a much higher voltage across the lamp ends to literally rip the electrons from the cold cathodes by a high electric field. They require much larger ballasts than preheat lamps in order to provide the required starting voltage.

    Rapid start lamps are a compromise between the other two types. Starting voltage is applied instantly, but so is preheating current for the cathodes by means of two small transformer windings in the ballast. The result is nearly instant starting but at a much lower voltage than an instant start lamp would require.

144 / ILLUMINATION

4. Q. Distinguish between standard output, high output (HO), and power groove (PG) fluorescent lamps. (Handbook - pages 8-94 and 8-95)

A. Standard output fluorescent lamps generally operate at a "loading" of about 10 watts per foot with a current of 430 ma. High output lamps run at 800 ma with 14 watts per foot, very high output and power groove lamps at 1,500 ma and 25 watts per foot. HO lamps provide about 35% more lumen output than standard lamps, VHO and PG lamps about 100% more. The price one pays for this increased output is a reduction in lumen per watt efficiency of 10% for HO lamps and 20% for VHO and PG lamps.

5. Q. Compare mercury, metal halide, and high pressure sodium high intensity discharge lamps. (Handbook - pages 8-33 to 8-37)

A. The mercury lamp was the forerunner of all high intensity discharge lamps. In the 400-watt size, it gives 22,500 initial lumens, has a life of 24,000 hours, and has an initial lamp efficiency of about 55 lumens per watt. It also has an undesirable color spectrum consisting of four spectral lines in the yellow-green, green, blue, and violet regions of the spectrum. It makes red and orange surfaces appear gray and is unkind to skin tones.

Over the years efforts have been made to improve the color characteristics of mercury lamps by adding a red-rich phosphor to the inside of the outer glass jacket. However, this eliminated the point-source virtue of the mercury lamp, making it unsuitable for use with a reflector.

The metal halide lamp is a mercury lamp to which halide salts of indium, thallium, scandium, and/or sodium have been added. These elements have characteristic spectral lines in positions of the spectrum where mercury does not and provide a lamp with a balanced color output. The 400-watt metal halide lamp provides 34,000 initial lumens at 12,000 hours life with an initial lamp efficiency of 85 lumens per watt.

The high pressure sodium lamp is the newest member of the HID lamp family. The low pressure sodium lamp was introduced in the 1930's but produced only the characteristic yellow sodium line at 590 nanometers. It was widely used in Europe for street lighting, but not in the U.S.

In the high pressure sodium lamp, the close packing of the sodium atoms produces a broad spectrum centered at 590 nm. The lamp gives a warm white light somewhat deficient in blues and violets. Nevertheless, its overall rendition of color is quite acceptable!

The high pressure sodium lamp is the most efficient light source yet devised. In the 400-watt size, it emits 50,000 initial lumens, has a life of 20,000 hours, and an initial lamp efficiency of 125 lumens per watt.

Chapter 10

ENGINEERING ECONOMY

In a capitalistic economy the success of engineering projects is usually measured in terms of financial efficiency. It is unlikely that a project will achieve maximum financial success unless it is properly planned and developed with concern for the technical, social, and financial requirements. The engineer is the one individual most likely to know the technical requirements of a project and he very frequently is required to make a study combining the technical and financial details of a project and thus provide an economic analysis upon which management can reach a decision.

Engineering economy refers to economic analysis of primarily engineering and technical projects. Since all engineering problems can be solved in more than one way, economy studies deal with the differences in economic results from the stated alternatives.

Engineering problems in economy usually involve a determination of what is economical over the long term, that is, over a considerable period of time. In these problems it is essential to recognize the time value of money and the existence of interest; interest represents money paid for use of borrowed money.

Economic analysis begins with interest problems, including ordinary simple interest, exact time and equations of values, and present value of a debt.

Economic studies require a knowledge of compound interest and amount, nominal and effective rates, and approximation of the interest rate and of the time. Analysis uses studies of periodic payment, term and rate of interest, amortization of a debt including depreciation and depletion.

Allocating the cost of plant and equipment is an important activity. The term plant and equipment refers to those long-lived assets acquired for use in the operation of the business (not intended for resale to customers). Plant and equipment include land, buildings, equipment-machinery, furniture and fixtures, office equipment and automobiles.

Plant assets, without the land, are of use to a company for only a limited number of years, since the cost of each plant and equipment asset can be allocated as an expense for the years in which there is use. In the accounting process this allocation is called <u>depreciation</u> which describes the gradual conversion of the cost of plant and equipment asset into expense. Depreciation in accounting refers to the allocation of the cost of a plant asset to the periods in which services are received from the asset.

There are several methods of calculating depreciation. A company need not use the same method of depreciation for all its various assets. The most commonly used methods are: straight-line; units-of-output; fixed percentage on declining balance; and sum-of-the digits.

In solving engineering economy problems consideration should be given to the <u>fixed input</u> (amount of money available), the <u>fixed output</u> or the minimizing of costs, and other variables not included above, i.e., maximized profit requirement. One important technique here is <u>present worth analysis</u> which determined the present value of future money receipts and disbursements. If the future income and costs are realized, then using a reasonable interest rate, the present worth of a business property can be calculated. In present worth technique careful consideration can be given to the time period covered by the analysis, i.e., useful life.

Engineering economy problems may also require an analysis of the annual cash flow which involves the standard accounting techniques. Engineering analysis is also concerned with "rate of return;" this is defined as the interest rate paid on the unpaid balance of a loan or the rate of return of interest paid on an investment, e.g., the re-payment schedule makes the unrecovered investment equal to zero at the end of the life of the investment.

Engineering economy uses a group of techniques for the systematic analysis of alternative decisions, including present worth, future worth, annual cost, rate of return and depreciation. The accompanying problems and solutions provide an illustration of these techniques; candidates not familiar with the above should review an engineering economy textbook.

<u>References</u>:

DeGarmo, E.P. and Canda, J.R. <u>Engineering Economy</u>, 5th edition. New York: The Macmillan Company, 1973.

Fabrycky, W.J. and Thuesen, G.J. <u>Economic Decision Analysis</u>. Englewood Cliffs, N.J.: Prentice-Hall Incorporated, 1974.

Grant, E.L. and Ireson, W.G. <u>Principles of Economy</u>, 5th edition. New York: The Ronald Press Company, 1970.

ENGINEERING ECONOMICS FORMULAS

Symbols used in interest formulas:

i represents an interest rate per interest period

n represents a number of interest periods

P represents a present sum of money

F represents a sum of money at the end of n periods from the present date that is equivalent to P with interest i

A represents the end-of-period payment or receipt in a uniform series continuing for the coming n periods, the entire series equivalent to P at interest rate i

G represents a uniform arithmetic gradient

Formulas

Single Payment Series:

Given P, to find F, $F = P(1 + i)^n$, $F = P(caf' - 1 - n)$,

$F = P(F/P)$ where $(1 + i)^n$ is called the single payment compound amount factor

Given F, to find P, $P = F \left[\dfrac{1}{(1+i)^n}\right]$, $P = F(pwf' - i - n)$,

$P = F(P/F)$ where $\dfrac{1}{(1+i)^n}$ is called the single payment present worth factor

150 / ENGINEERING ECONOMY

Uniform Annual Series

Given F, to find A, $A = F \left[\dfrac{1}{(1+i)^n - 1} \right]$, $A = F(sff - i - n)$;

$A = F(A/F)$, where $\dfrac{1}{[(1+i)^n - 1]}$ is called the sinking fund factor

Given P, to find A, $A = P \dfrac{[i(1+i)^n]}{[(1+i)^n - 1]}$ or

$A = P \dfrac{[i]}{[(1+i)^n - 1 + i]}$, $A + P(crf - i - n)$; $A = P(A/P)$

where $\dfrac{[i(1+i)^n]}{[(1+i)^n - 1]}$ and $\dfrac{[i]}{[(1+i)^n - 1 + i]}$ are capital recovery factors

Given A, to find F, $F = A \dfrac{[(1+i)^n - 1]}{[i]}$, $F = A(caf - i - n)$;

$F = A(F/A)$ where $\dfrac{[(1+i)^n - 1]}{[i]}$ is called the uniform series compound amount factor

Given A, to find P, $P = A \dfrac{[(1+i)^n - 1]}{[i(1+i)^n]}$, $P = A(pwf - i - n)$;

$P = A(P/A)$ or $P = A \dfrac{[1]}{\left[\dfrac{1}{(1+i)^n - 1} + 1\right]}$ are called the

uniform series present worth factors

Uniform Gradient Series

Given G, to find A,

$$A = \frac{G}{i} - \frac{nG}{1}\left[\frac{1}{(1+i)^n - 1}\right], \quad A = G(gf - i - n),$$

$$A = G(A/G)$$

Given G, to find P,

$$P = A(pwf - i - n) = G(gf - i - n)(pwf - i - n)$$

$$= G(gpwf - i - n)$$

$$P = G(P/G)$$

Identities

$$(caf' - i - n) = \frac{1}{(pwf' - i - n)} \; ; \quad F/P = 1/(P/F)$$

$$(sff - i - n) = \frac{1}{(caf - i - n)} \; ; \quad A/F = 1/(F/A)$$

$$(crf - i - n) = \frac{1}{(pwf - i - n)} \; ; \quad A/P = 1/(P/A)$$

$$(crf - i - n) = (sff - i - n) + 1 \; ; \quad A/P = A/F + 1$$

Effective Interest Rate

$$i = \left(1 + \frac{r}{m}\right)^m - 1 \quad \text{where } i = \text{effective annual interest rate}$$

$$r = \text{nominal annual interest rate}$$

$$m = \text{compounding period}$$

Finding Unknown Interest Rates:

$$F = P(1+i)^n \quad , \quad 1+i = \sqrt[n]{F/P}$$

$$i = \sqrt[n]{F/P} - 1 \qquad \text{Solution is by logarithms.}$$

SIMPLE INTEREST

Notation:

n = number of time periods
i = periodic interest rate expressed as a decimal
P = principal
FV = future value
I = interest amount

Interest amount

Formula: $$I = P \cdot n \cdot i$$

Example: Find the interest payment due of $1500 on a 360-day basis at 6% simple interest for 200 days.

Answer: $50.00 (Note: $i = 0.06/360$)

Number of time periods

Formula: $$n = \frac{FV-P}{P \cdot i}$$

Example: How long does it take to yield $1950 at 6% simple interest if the present value is $1500?

Answer: 5 years (Note: $i = 0.06$)

Interest rate

Formula: $$i = \frac{FV-P}{P \cdot n}$$

Example: Find the simple interest rate if $1500 invested today will amount to $1950 in 5 years.

Answer: 0.06 (6%)

Present value

Formula:
$$P = \frac{FV}{1+ni}$$

Example: What sum invested today at 6% simple interest will amount to $1950 in 5 years?

Answer: $1500 (Note: i = 0.06)

Future value

Formula:
$$FV = P(1 + ni)$$

Example: Find the future value of $1500 invested at 6% simple interest for 5 years.

Answer: $1950.00 (Note: i = 0.06)

COMPOUND INTEREST

Notation:

n = number of time periods
i = periodic interest rate expressed as a decimal
P = principal
FV = future value
I = interest amount

Interest amount

Formula:
$$n = \frac{\ln(\frac{FV}{P})}{\ln(1+i)}$$

Example: How long does it take to yield $2007.34 at 6% compounded annually if the principal is $1500?

Answer: 5 years (Note: $i = 0.06$)

Rate of return

Formula:
$$i = (\frac{FV}{P})^{1/n} - 1$$

Example: Find the rate of return if $1500 invested today compounded annually will amount to $2007.34 in 5 years?

Answer: 0.06 (6%)

Present value

Formula:
$$P = \frac{FV}{(1+i)^n}$$

Example: What sum invested today at 6% compounded annually will amount to $2007.34 in 5 years?

Answer: $1500.00 (Note: i = 0.06)

Future value

Formula: $$FV = P(1 + i)^n$$

Example: Find the future amount of $1500 invested at 6% compounded annually for 5 years.

Answer: $2007.34 (Note: i = 0.06)

Compound continuously

Formula: $$FV = P \cdot e^{in}$$

Example: Determine the value of $50 deposited at 6% for 5 years, compounded continuously.

Answer: $67.49 (Note: i = 0.06)

Nominal rate converted to effective annual rate

Formula: $$\text{Effective rate} = (1 + i)^n - 1$$

Example: What is the effective annual rate of interest if the nominal (annual) rate of 12% is compounded quarterly?
(n = 4, i = 0.12/4)

Answer: 0.1255 (12.55%)

PERIODIC PAYMENTS

Notations:

n = number of payments
i = periodic interest rate expressed as a decimal
PMT = payment
P = principal

Number of time periods

Formula:

$$n = \log_{1+i}\left(\frac{1}{1 - \frac{P \cdot i}{PMT}}\right)$$

Example: How many payments does it take to pay off a loan of $4000 at 9.5% annual rate, with payments close to $150 per month?

Answer: 30.07 payments (Note: $i = 0.095/12$)

Interest rate

Formula:

$$\text{Monthly interest rate } i = \frac{PMT\; 1 - (\frac{1}{1+i})^n}{1}$$

Annual interest rate = monthly rate x 12

Example: If n = 360, monthly payment PMT = 179.86, P = 30000, find the annual interest rate.

Answer: 6.00%

Payment amount

Formula:

$$PMT = \frac{P \cdot i}{1 - (1+i)^{-n}}$$

158 / ENGINEERING ECONOMY

Example: To pay off a loan of $4000 at 9.5% interest in 30 months, what monthly payment is required?

Answer: $150.32 (Note: i = 0.095/12)

Present value

Formula:
$$P = PMT \frac{1 - (1+i)^{-n}}{i}$$

Example: A person is willing to pay $150 per month for 30 months for a loan at 9.5%, how much can be borrowed?

Answer: $3991.55 (Note: i = 0.095/12)

Accumulated interest

Formula: The interest paid from payment j to payment k is

$$I_{j-k} = PMT \left[k-j - \frac{(1+i)^{k-n}}{i}(1 - (1+i)^{j-k}) \right]$$

Compute the monthly payment, PMT, by the formula given above under "Payment Amount".

Example: Consider a business property costing $30.000 with a mortgage life of 30 years at 8% yearly interest. Find the interest paid on the first 36 montly payments (i = 0.08/12, j = 0, k = 36, n = 360).

Answer: PMT = $220.13
I_{0-36} = $7108.72

Remaining balance

Formula: The remaining balance at payment k (k = 1, 2, 3, ..., n) is

$$P_k = \frac{PMT}{i} \left[1 - (1+i)^{k-n} \right]$$

Example: Using the previous example, find the remaining balance at payment 36.

Answer: $29184.13

ENGINEERING ECONOMY / 159

Interest rebate (Rule of 78's)

Formula: F = finance charge

$$I_k = \text{interest charged at month } k = \frac{2(n-k+1)}{n(n+1)} F$$

$$\text{rebate} = \frac{(n-k)I_k}{2}$$

Example: A 30 month, $1000 loan having a finance charge of $180.00 is being repaid at $39.33 per month. What is the interest portion of the 25th payment? What is the interest rebate at that point?

Answers: Interest portion of the 25th payment = $2.32
Rebate = $5.81

DEPRECIATION AMORTIZATION

Straight line depreciation

Formulas:
$$D = \frac{PV}{n}$$

$$B_k = PV - kD$$

where PV = original value of asset (less salvage value)
n = lifetime periods of asset
D = each year's depreciation
B_k = book value at time period k

Example: Equipment have a value of $2100 and a life expectancy of six years. Using the straight line method, what is the amount of depreciation and what is the book value after two years?

Answers: D = $350.00
B_2 = $1400.00

Variable rate declining balance

Formulas:
$$D_k = PV \cdot \frac{R}{n} (1 - \frac{R}{n})^{k-1}$$

$$B_k = PV(1 - \frac{R}{n})^k$$

where PV = original value of asset
n = lifetime periods of asset
R = depreciation rate (given by user)
D_k = depreciation at time period k
B_k = book value at time period k

Example: Equipment has a value of $2500 and a life expectancy of six years. Use the double declining balance method (R = 2) to find the amount of depreciation and book value after four years.

Answers: $B_4 = \$493.83$
$D_4 = \$246.91$

Sum of the year's digits depreciation

Formula:
$$D_k = \frac{2(n - k + 1)}{n(n + 1)} PV$$

$$B_k = S + (n - k)D_k/2$$

where PV = original value of asset
n = life time periods of asset
S = salvage value
D_k = depreciation at time period k
B_k = book value at time period k

Example: Equipment has a value (less salvage value - $800) of $2100 and a life expectancy of 6 years. Using the SOD method, what is the amount of depreciation and what is the book value after 2 years?

Answers: $D_2 = \$500.00$
$B_2 = \$1800.00$

Diminishing balance depreciation

Formulas:
$$D_k = PV_{k-1} \left[1 - \left(\frac{S}{PV_o}\right)^{1/n} \right]$$

$$PV_k = PV_{k-1} - D_k$$

where PV_o = original value of asset
S = salvage value (>0)
PV_k = book value at time period k (k = 1, 2,...n)
D_k = depreciation at time period k

Example: Equipment has a value of $2500, a salvage value of $400, and a life expectancy of six years. Find the amount of depreciation and book value for each of the first three years by using the diminishing balance method.

Answers: $D_1 = \$657.98$

162 / ENGINEERING ECONOMY

Answers cont'd: $PV_1 = \$1842.02$
$D_2 = \$484.81$
$PV_2 = \$1357.21$
$D_2 = \$357.21$
$PV_3 = \$1000.00$

ENGINEERING ECONOMY / 163

ADDITIONAL ENGINEERING ECONOMICS

Example of Benefit-Cost Analysis

Four alternatives are compared as follows, assuming a 30 year life for each project, and i = 10%.

	W	X	Y	Z
Initial investment	200,000	250,000	300,000	350,000
C.R. = P(crf - i - 30) = P(A/P - i - 30) = P(0.10608)	-21,216	-26,520	-31,824	-37,128
Annual costs and disbenefits	-7,000	-7,500	-8,000	-8,500
Annual increased benefits	29,000	35,000	40,000	45,000
Total annual costs and disbenefits	-28,216	-34,020	-39,824	-45,628
B - C	+784	+980	+176	-628
B/C	1.028	1.029	1.004	0.986

Z is eliminated because its B/C is less than 1.0. W, X, and Y all seem favorable since B/C is greater than 1.0. However, for a project to be acceptable the B/C ratio for its increased investment over the project with the next smaller investment must also be greater than 1.0. Then compare benefit-costs of the extra investments as follows:

	X-W	Y-W	Y-X
Annual benefits	6,000	11,000	5,000
Total annual costs and disbenefits	-5,804	-11,603	-5,804
B - C	+196	-608	-804
B/C	1.034	0.948	0.861

Since the B/C ratios for the additional investments of Y-W and Y-X are both less than 1.0, plan Y is eliminated, and since the B/C ratio for the additional investment X-W is greater than 1.0, X is preferable to W. This may be rationalized as follows:

B/C for X-W is greater than 1.0, therefore X is preferred to W;

B/C for Y-W is less than 1.0, therefore W is preferred to Y;

B/C for Y-X is less than 1.0, therefore X is preferred to Y;

Then X is the preferred plan.

ECONOMIC ANALYSIS PROBLEMS AND SOLUTIONS

Example 1

When business property is bought by a sequence of partial payments, the <u>buyers equity</u> in the property at any time is that of the price of the property which the company has paid. At the same time, the <u>seller's equity</u> is that part of the price of the property which remains to be paid, that is, the outstanding principal at the time.

ABX Company buys property with a small equipment storage structure for $35,000, and pays $6,000 down and mortgages the balance with interest at 12% compounded monthly by equal payments at the end of each month for the next 10 years.

1. What is the ABX Company's monthly payment?
2. What is ABX Company's equity in the structure just after making the 30th payment?
3. What is the seller's equity after 150 months?
4. How much money is ABX Company paying in interest for the twenty-year period?
5. If ABX Company made an initial down payment of $18,000, how much money would the company save on the payment?

Answers

1. Solution: Payment = R = $29{,}000 \dfrac{1}{a_{\overline{240}|.01}} = \dfrac{29000}{90.81} = \$3\ .34$

2. Solution: $319.34\, a_{\overline{190}|.01} = \$27112.09 \qquad 35000 - 27112.09 = \$7887.91$

3. Solution: $319.34\, a_{\overline{90}|.01} = \$18892.40$

166 / ENGINEERING ECONOMY

4. Solution $29000(1 + .01)^{240} = \$315,859$

5. Solution: $17000 \dfrac{1}{a_{\overline{240}|.01}} = \$187.18 \qquad 319.34 - 187.18 = \$132.16$

Example 2

A 20 year annuity bond for $30,000.00, with interest at 8% compounded semiannually, is to be paid off in 40 equal semiannual installments, the first due 6 months from today.

1. The periodic installment is?

2. Find the purchase price at the end of the tenth year to earn 6% compounded semiannually.

3. If the interest was 8% compounded quarterly, what is the periodic installment?

4. What would the difference be in the installment payments of Question 1 and Question 3 for a 6 month period?

5. Find the purchase price at the end of the fifth year to earn 7% compounded quarterly using the figures from Question 3.

Answers

1. Solution: $30000 \dfrac{1}{a_{\overline{40}|.04}} = \dfrac{30000}{19.7927} = \$1515.71$

2. Solution: Purchaser is buying the right to collect the remaining 20 installments

   $p = 1515.71 \dfrac{}{a_{\overline{20}|.03}} = 1515.71(14.8774) = \$22,549.82$

3. Solution: $30000 \dfrac{1}{a_{\overline{80}|.02}} = \dfrac{30000}{39.7445} = \$754.82$

4. Solution: $1515.71 - 2(754.82) = 6.07$ less with installment plan in question 3.

5. Solution: Purchaser is buying the right to collect the remaining 60 payments

$$p = \frac{754.82}{a_{\overline{60}|.0175}} = \$27,901$$

Example 3

A consulting engineer decides to set up a retirement fund for himself. At the end of the current year he will be 30 years old. He plans to retire when he reaches 65. His basic plan is to make uniform deposits at the end of each year, starting with next year, to this retirement fund. He can expect to earn 9 per cent annual interest, compounded annually, on his money.

1. If he were to deposit $1,500 at the end of each of the next 15 years, the value of these deposits at the time of his retirement would be?

2. Suppose the engineer deposits $1,000 at the end of each year for the next 12 years and then suspends payments for 6 years because of serious illness. The amount in his retirement fund at the end of 18 years would be?

3. Suppose the engineer was unable to make any payments in the first 10 years as planned. If he then made payments at the end of each year until retirement, so that he had $50,000 at retirement, each of those payments would have to be?

4. The amount the engineer must deposit at the end of the current year to give him $82,000 at retirement is?

5. If the engineer had been able to receive 6 per cent annual interest, compounded semiannually, on an original deposit of $15,000, its value at retirement age would be?

6. Suppose that at the end of 21 years of somewhat erratic payments the engineer has $49,000 in his fund. What lump sum could he have deposited initially to obtain the same amount?

7. If the engineer deposited $500 into his account each year for the next 15 years, how much would he have to add in a lump sum at the end of that 15 years so that he would have $150,000 at retirement age?

168 / ENGINEERING ECONOMY

8. Assume 9 per cent annual interest. Plan A requires a single investment of $2,000. Plan B requires uniform savings of $230 per year. If the engineer wants to have $50,000 in his retirement fund after 35 years, which plan comes the nearest to the $50,000 goal?

9. If all the engineer could save was $400 a year, how much interest would he receive a year when he retires?

10. If the economy suddenly took a turn for the worse and all he could get for his money was 2% annual interest, how much would an annual savings of $50 accumulate to at his retirement age?

Answers

1. Solution: $\dfrac{1,500(1 + .09)^{15}}{.09} = 44,041$ at end of 15 years

   Single payment for 20 years $44,041(1 + .09)^{20} = \$246,825$

2. Solution: F/A $\quad 1000 \dfrac{(1+.09)^{12}-1}{.09(1 + .09)^{12}} = 7160.9$

   F/P $\quad 7160.9 (1+.09)^{6} = \$12,009$

3. Solution: $50,000 = p \dfrac{(1 + .09)^{25}}{.09} = \$1590$

4. Solution: $82,000 = p(1 + .09)^{35} \qquad p = \$4,016$

5. Solution: $15,000(1 + .06)^{70} = 886,144$

6. Solution: $p(1 + .09)^{18} = 49,000 \quad p = 10,387$

7. Solution: The value of $500 deposited each year for 15 years
   $500\dfrac{(1 + .09)^{15}-1}{.09} = \$14,680$

   f/P for 20 years
   $(14,680 + x)(1 + .09)^{20} = 150,000$

Solution 7 cont'd

$$(14,680 + x)(5.604) = 150,000$$

$$82267 + 5.604x = 67,733 \qquad x = \$12,086$$

8. Solution: $230 \dfrac{(1 + .09)^{35} - 1}{.09} = \$49,614 \qquad \$380$ from goal

9. Solution: $400 \dfrac{(1 + .09)^{35} - 1}{.09} = \$86,285 \qquad 86,285 \times .09 = \$7,765$

10. Solution: $50 \dfrac{(1 + .02)^{35} - 1}{.02} = \$12,500$

170 / ENGINEERING ECONOMY

Example 4

The sum of $19,000 will be needed at the end of 9 years. Money can be invested to earn 7½% effective.

1) If the person wants to set aside the same amount each year, how much must he invest each year?

2) The amount in the fund at the end of the sixth year is?

3) What is the amount of interest (cumulative) at the end of the fifth year?

4) If the person wanted to make one lump investment initially, how much would he be investing?

5) Due to illness, the person could not make the last investment. What would be the amount in the fund at the end of 9 years?

Answers

1) Solution:

$$19,000 = R \, s_{\overline{9}|\, 7\frac{1}{2}\%}$$

$$R = 19,000 \, \frac{1}{s_{\overline{9}|\, 0.75}} = 19,000 \times .0817 = \$1,552.30$$

2) Solution:

$$1,552.30 \times s_{\overline{6}|\, 0.75}$$

$$1,552.30 \times 7.2463 = \$11,248.55$$

3) Solution:

value of fund at end of fifth year

$$1,552.30 \times s_{\overline{5}|\, 0.75} = \$9,016.61$$

ENGINEERING ECONOMY / 171

3) Solution: (continued)

   amount paid in

   $5 \times 1{,}552.30 = \$7{,}761.50$

   interest = $9{,}016.61 - 7{,}761.50$

   interest = $\$1{,}255.11$

4) Solution:

   $19{,}000 = R(1 + i)^n$

   $19{,}000 = R(1.075)^9$

   $R = 19{,}000/(1.075)^9 = \$9{,}910.15$

5) Solution:

   value of fund at end of 8 years

   $1{,}552.30 \times s_{\overline{8}|.075} = \$16{,}220.48$

   for last year

   $16{,}220.48 \times (1 + .075) = \$17{,}437.00$

Example 5

A $20,000 annuity bond will be redeemed, principal and interest, at $j_1 = 4\%$, in 15 equal annual installments.

1) What will the installment payments be?

2) What will a prospective buyer offer for the bond if he wishes to realize an effective rate of 6%?

3) Same as question #2, except effective rate of 5%?

4) A comapny buys a piece of machinery for $15,000. It costs $550 per year for maintenance and brings in $1,350 per year rental fees. At the end of 8 years, it is sold for $2,500 salvage value. What is the economic profit at the end of 8 years?

5) If $15,000 is invested at an interest rate of 8% compounded annually, which investment plan, #4 or #5, is better and by how much?

Answers

1) Solution:

$$20{,}000 = R \, a_{\overline{15}|\,4\%}$$

$$R = \frac{20{,}000}{a_{\overline{15}|\,4\%}} = \frac{20{,}000}{11.1183} = \$1{,}798.83$$

2) Solution:

$$p = 1{,}798.83 \times a_{\overline{15}|\,6\%}$$

$$p = \$17{,}470.60$$

3) Solution:

$$p = 1{,}798.83 \times a_{\overline{15}|5\%}$$

$$p = \$18{,}671.14$$

4) Solution:

initial cost − $15,000
annual cost − $550 per year x 8 years = $4,400

total cost = $19,400 = C

Income = $1,350/year x 8 years = $10,800
Salvage =   2,500
                                                  ─────
                                                  $13,300

p = C − I

p = $19,400 − $13,300

p = $6,100

5) Solution:

$$\$15{,}000(1 + .08)^8 = \$27{,}764$$

$$\$27{,}764 - \$15{,}000 = \$12{,}764 \text{ profit}$$

Part III

Second Sample Principles and Practice of Electrical Engineering Examination

MORNING SECTION

You will have four hours in which to work this test. Your score will be directly proportional to the number of problems you solve correctly through four (4). Each correct solution counts ten points. The maximum possible score for this part of the examination is 40 points. Partial credit for partially correct solutions will be given.

Work four of the problems according to instructions. Do not submit solutions or partial solutions for more than four problems. Indicate the problems which you have solved.

You may work only one engineering economy problem. When you have completed this portion of the P&P examination in the required time limit, you should check your solutions with the answers at the end of this examination. For additional practice you are encouraged to work the other problems under the time limit.

Since you want the maximum points available, you should remember that the examiner who assigns these points must make his judgment based only on what has been written down during the examination. It is important for you to be reasonably neat in your work and write down any assumption that you consider necessary to allow you to solve the problem properly and to provide sufficient rationale so that the examiner can judge your reasoning. Assumptions should follow the logic and requirements of the problem.

You are advised to use your time effectively.

SECOND SAMPLE P & P/EE EXAMINATION

1. A sports arena is planned in a growing metropolis. The administrative offices and entrance halls to the arena face on a street that is expected to show great commerical progress. Two plans are offered for the construction of this part of the arena. Plan I is to construct a three-story building with provision for adding 12 stories at a future date. The cost of this initial building, designed to take the future loads, will be $3,100,000. In 10 years the stories can be added at a cost of $11,200,000. The life of the entire structure, including the arena is expected to be 35 years with zero salvage.

Plan II is to construct for $1,110,000 a three-story building which will be demolished in 10 years and replaced by a 15-story structure at a cost of $15,300,000.

The maintenance of Plan I will be $40,000 and of Plan II, $35,000 for the first 10 years. After that they will be alike. The property taxes will be 4% of the first costs of the structure. The minimum required rate of return will be 20%. Make your recommendation based on a present-worth analysis.

2. The circuit shown can be either a NAND or NOR gate, depending on the resistor values in the input circuit. Determine the values of R_1, R_2, R_3 and R_B for the conditions given and assuming the circuit is a positive Logic NAND gate.

   Logical 1 = +5V
   Logical 0 = 0.4V

   V_{BB} = -5V V_{CC} = 5V
   $V_{BE_{sat}}$ = 0.2V $V_{CE_{sat}}$ = 0.4V
   $I_{B_{sat}}$ = 100μa $I_{C_{sat}}$ = 4.6ma
   R_L = 1K $R_1 = R_2 = R_3$

NAND and NOR Truth Tables

		NAND			NOR	
A	B	C	Output ($\overline{ABC}$)	Output	$\overline{(A + B + C)}$	
0	0	0	1	1		
0	0	1	1	0		
0	1	0	1	0		
0	1	1	1	0		
1	0	0	1	0		
1	0	1	1	0		
1	1	0	1	0		
1	1	1	0	0		

Resistor-transistor logic (RTL) NAND or NOR gate. Output $\overline{ABC}$ or $\overline{A + B + C}$, depending on the values of $R_{1,2,3}$ and R_B.

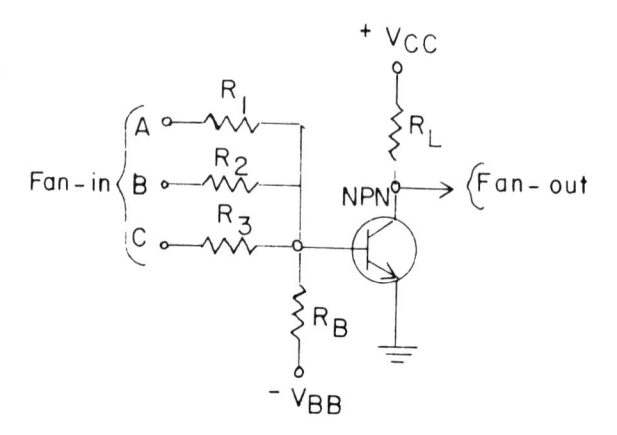

3. A 3-phase Y-connected 220-volt (line to line) 10-hp 60-cps 6-pole induction motor has the following constants in ohms per phase referred to the stator:

$$r_1 = 0.294 \qquad r_2 = 0.144$$
$$x_1 = 0.503 \qquad x_2 = 0.209 \qquad x_\phi = 13.25$$

The total friction, windage, and core losses may be assumed to be constant at 403 watts, independent of load.

For a slip of 2.00 per cent, compute the speed, output torque and power, stator current, power factor, and efficiency when the motor is operated at rated voltage and frequency.

4.

For the R-C coupled, N-channel FET amplifier shown, the following signal and d-c parameters are available on T_1 and T_2:

g_f = 2000 μmho r_d = 100 KΩ

C_{gs} = 2 pF C_{dg} = 1 pF C_w is negligible

V_{DS} = 10 V V_{GS} = -3 V BV_{GDS} = 25 V

I_D = 2 mA I_{GSS} = 0.1 μA

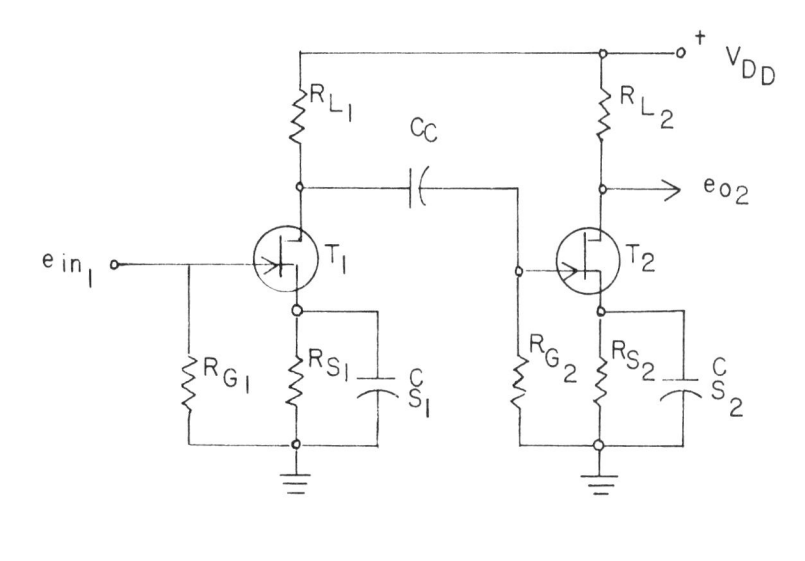

Find:

a) $R_{S_{1,2}}$

b) V_{DD}

c) $R_{L_{1,2}}$

d) $R_{G_{1,2(max)}}$

e) Overall midband gain

f) C_{in_2} and f_T

g) Upper breakpoint-frequency f_2

h) Lower breakpoint-frequency f_1, if C_C = 0.01μF

i) $C_{S_{1,2}}$ for adequate R_S bypass at f_1

j) Minimum d-c WV of C_C and C_S

180 / SECOND SAMPLE P & P/EE EXAMINATION

5. Ten decibels of negative feedback are required on a single-stage transistor having open-loop gains of $A_i = 40$ and $A_v = -50$. The current feedback principle of the figure shown is to be employed on the stage.
Using these data, complete the following:

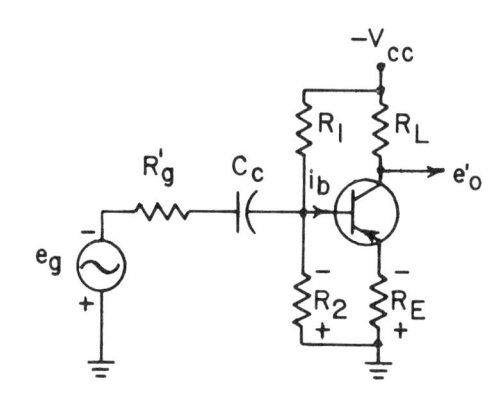

a) What will be the necessary β (feedback ratio) value?

b) If $R_L = 2K\Omega$, what will be the value of R_E?

c) Assuming an f_T of 800 MHz of the stage and a 6dB per octave roll-off, what will be the upper breakpoint frequency with and without feedback?

d) If the open-loop input resistance is $r_{in} = 1K\Omega$, what is r'_{in} (the closed-loop input resistance)?

e) Determine the V_{CC} supply value when V_{CE} is 5V, and $I_C = 4mA$.

f) Determine R_B when $I_B = 50\mu A$ and $V_{BE} = 0.2V$

6. A display room is 54 ft. by 30 ft. with a 13-ft. ceiling and has display counters along each wall as well as lengthwise along the long axis of the room. The reflection factor of the ceiling is 70%, and that of the walls is 50%. Design a lighting system using a 4-lamp luminaire having luminous side panels. The luminaire shielding is 35 degrees lengthwise and 35 degrees crosswise. The lamps are to be 48-in. T-12 standard cool white lamps. Maintenance is medium. The in-service foot-candle level is to be 100 ft. -C. Assume the luminaire data given below to apply. Make a rough sketch of layout of luminaire (scale drawing not required) with dimensions to show spacing, etc.

Distribution	Spacing not to exceed	Maintenance Factor	
44 ↑↓ 43	1.2xM.H.	Good	.70
		Med.	.65
		Poor	.60

	Ceiling	80%			70%			50%		
	Walls									
	Room									
	Index	50%	30%	10%	50%	30%	10%	50%	30%	10%
		Coefficients of Utilization								
J		.27	.21	.18	.26	.21	.18	.24	.20	.17
I		.35	.28	.24	.33	.27	.24	.30	.26	.23
H		.39	.33	.28	.37	.32	.28	.34	.30	.26
G		.46	.40	.35	.44	.38	.33	.40	.35	.31
F		.51	.44	.39	.48	.42	.38	.43	.38	.35
E		.57	.51	.46	.54	.48	.44	.48	.43	.40
D		.62	.56	.52	.58	.54	.49	.52	.48	.44
C		.65	.60	.55	.61	.57	.53	.54	.50	.47
B		.70	.65	.61	.65	.62	.58	.57	.54	.52
A		.73	.69	.65	.68	.65	.62	.59	.57	.54

7. A common emitter PNP transistor is designed to be operated in an R_E stability-biased circuit with a stability factor of 6 and an R_E of $0.5R_L$. The lowest frequency to be amplified to $80H_z$. Given the following

$V_{CE} = -5V$ $I_B = -50\mu A$ $I_{CO} = -2\mu A$

$I_C = -2mA$ $V_{BE} = -0.18V$ $R_L = 2K\Omega$

$h_{ie} = 1.5K\Omega$ $h_{fe} = 40$

$h_{oe} = 40.10^{-6}$ mho $h_{re} = 3.10^{-4}$

$V_{be} = 30mV$ p-to-p $i_b = 20\mu A$ p-to-p

Determine:
- a) R_E
- b) V_{CC}
- c) R_1
- d) A_v
- e) The optimum value of R_L having the indicated h-parameters.

8. Find the trigonometric Fourier series for the waveform shown and sketch the line spectrum.

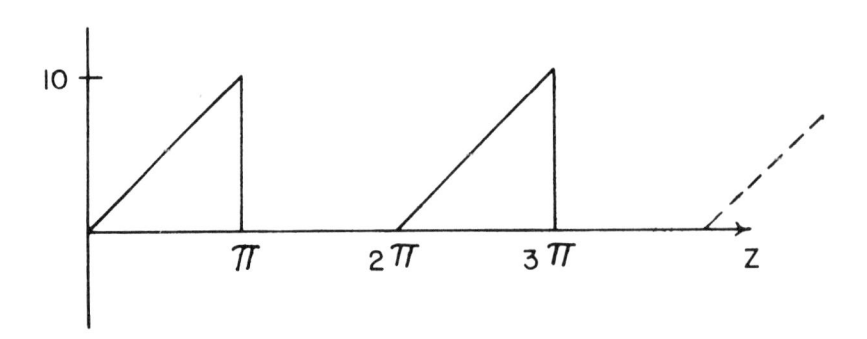

9. In the parallel circuit shown below, the total power is 1100 watts. Find the power in each resistor and the reading on the ammeter.

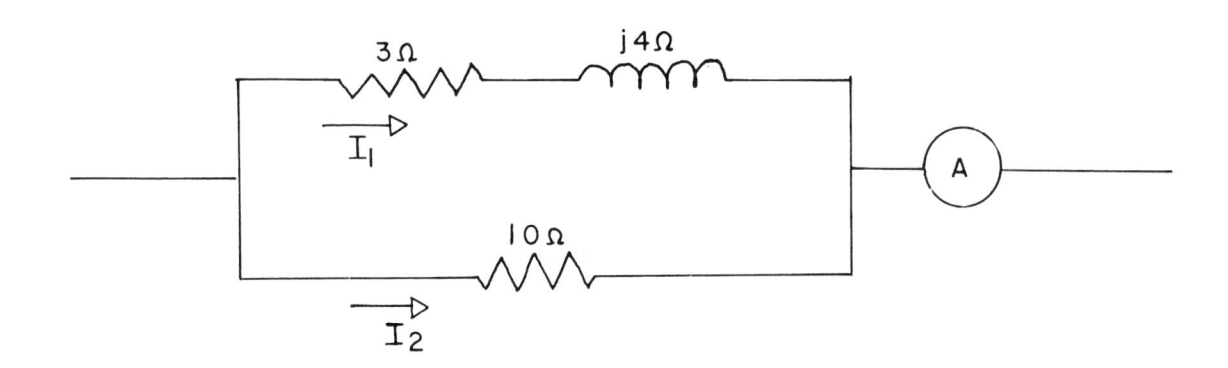

Solution 1:

$$PW_I = 3{,}100{,}000 + 164{,}000 \underset{.20-10}{(uspwf)}^{4.1925} + 11{,}200{,}000 \underset{10}{(sppwf)}^{.16151}$$

$$+ 572{,}000 \underset{25}{(uspwf)}^{4.9476} \underset{10}{(sppwf)}^{.16151} = \$6{,}053{,}000$$

$$PW_{II} = 1{,}100{,}000 + 79{,}000 \underset{10}{(uspwf)}^{4.1925} + 15{,}300{,}000 \underset{10}{(sppwf)}^{.16151}$$

$$+ 612{,}000 \underset{25}{(uspwf)}^{4.9476} \underset{10}{(uspwf)}^{.16151} = \underline{\$4{,}389{,}000}$$

Solution 2:

$$R_{in_{sat}} = \frac{V_{BE_{sat}}}{I_{B_{sat}}} = \frac{0.2}{100\mu a} = 2K$$

Critical state 110, 101, 011

$$i_1 + i_2 - i_3 - i_4 - i_5 = 0$$

'ON' Transistor

$$\frac{V_D - V_{BB}}{R_B} + \frac{V_D - 0}{R_{in_{sat}}} - \frac{V_A - V_D}{R_1} - \frac{V_B - V_D}{R_2} - \frac{V_C - V_D}{R_3} = 0$$

Since $R_1 = R_2 = R_3$ and $V_A = V_B = V_C$

$$\frac{V_D - V_{BB}}{R_B} + \frac{V_D}{R_{in_{sat}}} - 3\frac{V_A - V_D}{R} = 0 \quad \text{Eq for 'on'}$$

$$i_1 - i_2 - i_3 - i_4 = 0$$

$$\frac{V_D - V_{BB}}{R_B} - \frac{V_A - V_D}{R_1} - \frac{V_B - V_D}{R_2} - \frac{V_C - V_D}{R_3} = 0$$

'OFF' Transistor

Solution 2 cont'd

$$\frac{V_D - V_{BB}}{R_B} - \frac{2(V_A - V_D)}{R} - \frac{V_C - V_D}{R} = 0$$

Eq for 'off'

$$\frac{0.2(-5)}{R_B} + \frac{.2}{2K} - \frac{3(5-.2)}{R} = 0$$

$$\frac{-.1-(-5)}{R_B} - \frac{2(5-(-.1))}{R} - \frac{.4-(-.1)}{R} = 0$$

$R = R_1 = R_2 = R_3 = 30.5K$

$R_B = 14K$

Solution 3:

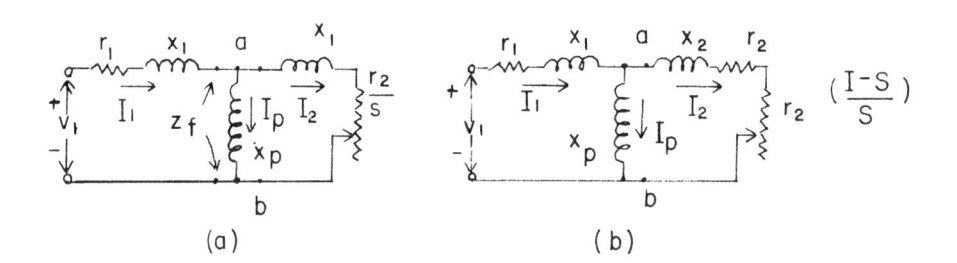

Equivalent Circuits

The impedance Z_f represents physically the per-phase impedance presented to the stator by the air-gap field, both the reflected effect of the rotor and the effect of the exciting current being included therein.

$$Z_f = R_f + jX_f = \frac{r_2}{S} + jx_2 \text{ in parallel with } jx_\phi$$

Substitution of numerical values gives, for s = 0.0200,

$R_f + jX_f = 5.41 + j3.11$
$r_1 + jx_1 = \underline{0.29 + j0.50}$
  Sum = $5.70 + j3.61 = 6.75\underline{/32.4}$ ohms

Applied voltage to neutral = $\frac{220}{\sqrt{3}}$ = 127 volts

Solution 3 cont'd

Stator current $I_1 = \dfrac{127}{6.75} = 18.8$ amp

Power factor $= \cos 32.4° = 0.844$

Synchronous speed $= \dfrac{2f}{p} = \dfrac{120}{6} = 20$ rev/sec, or 1,200 rpm

$\omega_s = 2\pi(20) = 125.6$ rad/sec

Rotor speed $= (1-s) \times$ (synchronous speed)
$= (0.98)(1,200) = 1,176$ rpm

$$P_{g1} = q_1 I_2^2 \dfrac{r_2}{s} = q_1 I_1^2 R_f$$

$$= (3)(18.8)^2(5.41) = 5,740 \text{ watts}$$

the internal power is

$P = (0.98)(5,740) = 5,630$ watts

Deducting losses of 403 watts gives

Output power $= 5,630 - 403 = 5,230$ watts, or 7.00 hp

Output torque $= \dfrac{\text{output power}}{\omega_{rotor}} = \dfrac{5,230}{(0.98)(125.6)} = 42.5$ newton-m, or 31.4 lb-ft

The efficiency is calculated from the losses.

Total stator copper loss $= (3)(18.8)^2(0.294)$	$= 312$ watts
Rotor copper loss $(0.0200)(5,740)$	$= 115$
Friction, windage, and core losses	$= 403$
Total losses	$= 830$ watts
Output	$= 5,230$
Input	$= 6,060$ watts

$\dfrac{\text{Losses}}{\text{Input}} = \dfrac{830}{6060} = 0.137$ Efficiency $= 1.000 - 0.137 = 0.863$

Solution 4:

a) $R_{S_{1,2}} = \dfrac{V_{GS}}{I_D} = \dfrac{3}{2ma} = 1.5K$

186 / SECOND SAMPLE P & P/EE EXAMINATION

Solution 4 cont'd

b) Let $V_{DD} = BV_{GDS} = 25$

c) $R_{L_{1,2}} = \dfrac{V_{DD}-V_{DS}}{I_D} - R_S = \dfrac{25-10}{2ma} - k.5\ K = 6K$

d) $R_{G_{1,2}(max)} = \dfrac{0.01 I_D R_S}{I_{GSS}} = \dfrac{.01 \times 3}{0.1 \mu a} = 0.3M$

e) $A_{v_2} = -g_f \left(\dfrac{r_d R_L}{r_d + R_L} \right) = -2 \times 10^{-3} \left(\dfrac{100K \times 6K}{100K + 6K} \right) = -11.3$

   where $R_{L_{T_1}} = R_{L_1} \parallel R_G = 6K \parallel .3M \sim 6K$

   and $A_{v_1} = A_{v_2} = -11.3$

   $A_v(\text{overall MF}) = A_{v_1} A_{v_2} = (-11.3)^2 = 128$

f) $C_{in_2} = C_{gs} + C_{ds}(1+|A_{v_2}|) = 2pf + kpf(1+11.3) = 14.3pf$

   $C_S = C_{in_2}$ since C_W is negligible

   $f_\tau = \dfrac{g_f}{2\pi C_S} = \dfrac{2 \times 10^{-3}}{2\pi \times 14.3pf} = 22.2 Mhz$

g) $f_2 = f_\tau / A_{v_1} = \dfrac{22.2 MHz}{11.3} = 1.96 Mhz$

h) $R_a = r_d \parallel R_{L_1} = 100K \parallel 6K = 5.66K$

   $f_1 = \dfrac{1}{2\pi C_C(R_A + R_G)} = \dfrac{0.159}{0.01 \mu f (5.66K + .3M)} = 53 Hz$

Solution 4 cont'd

i) $x_{CS} = 0.1 R_S$ @ $f_1 = 0.1 \times 1.5K = 150$ @ 53 hz

$$C_S = \frac{0.159}{f_1 X_{CS}} = \frac{0.159}{53 \times 150} = 20.35 \mu f$$

j) C_C dc-WV is V_{DD} (minimum)

C_S dc-WV is $|V_{GS}|$ (minimum)

It is the practice to use a dc-WV of 2 times the minimum.

Solution 5

a) db = $20 \log (1-A_v \beta) = 10$

$1 - A_v \beta = 3.16$

$\beta = \frac{2.16}{-A_v} = \frac{2.16}{50} = 0.043$

b) $\beta = \frac{R_E}{R_L}$, $R_E = \beta R_L = 0.043 \times 2K = 86$

c) $f\tau = A_i BW = A_i f_2$

$$f_2 = \frac{f\tau}{A_i} = \frac{800 MHz}{40} = 20 MHz$$

$f'_2 = f_2 (1-A_v \beta) = 20 MHz \times 3.16 = 63 MHz$

d) $r'_{in} = r_{in} (1-A_v \beta) = 1K \times 3.16 = 3.16K$

e) $V_{CC} = I_C (R_L + R_E) + V_{CE} = 4mA(2K+86) + 5 = 13.34$

f) $V_{RE} \simeq I_C R_E = 4ma \times 86 = 0.34V$

$$R_B = \frac{V_{CC} - (V_{BE} + V_{RE})}{I_B} = \frac{13.34 - (0.2 + .34)}{50 \mu a} = 256K$$

Solution 6:

Refer to <u>IES Lighting Handbook,</u> 2nd. ED., 1952, Table 9-1

If it is assumed that the lights are to be dropped 2 to 3 feet below the ceiling, Table 9-1 indicates the room index should be D. If lights are to be ceiling mounted, Table 9-1 indicates a room index interpolated between E and F. Similar interpolation of the coefficient of utilization table supplied with the problem indicates a coefficient of utilization of approximately 0.50.

$$\text{No. of luminaires} = \frac{\text{(in-service foot-candle rating)}}{\text{(Lumens/lamp)(lamps.luminaire)}}$$

$$\times \frac{\text{(room area)}}{\text{(Coeff. of util.)(Maint. factor)}}$$

$$= \frac{100(54 \times 30)}{2650(4)(0.50)(0.65)} = 47.0$$

Suggests 7 rows of 7 luminaires/row (49 luminaires)

In-service foot-candles = $\frac{49}{47.0}$ (100) = 104

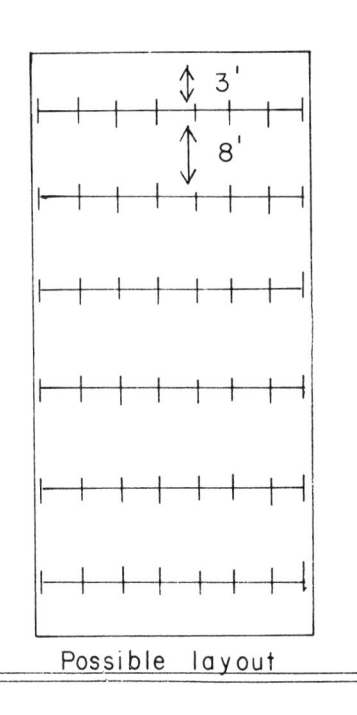

Possible layout

Solution 7:

a) $R_E = 0.5 R_L = 0.5 \times 2K = 1K$

b) $V_{CC} = V_{CE} + I_C R_{L_{DC}} = 5V + 2ma \times 3K = 11V$

   where $R_{L_{DC}} = R_E + R_L = (1K + 2K) = 3K$

c) $R_2 \doteq (S-1) R_E = (6-1) 1K = 5K$

   $V_{RE} = I_C R_E = 2ma \times 1K = 2V$

   $V_{R_2} = V_{RE} + V_{BE} = (2 + .18) = 2.18V$

   $I_{R_2} = \dfrac{V_{R_2}}{R_2} = \dfrac{2.18}{5K} = 0.43ma$

   $I_{R_1} = I_B + I_{R_2} = (.05 + 0.43) = 0.48ma$

   $V_{R_1} = V_{CC} - V_{R_2} = (11 - 2.18) = 8.82V$

   $R_1 = \dfrac{V_{R_1}}{I_{R_1}} = \dfrac{8.82}{0.48ma} = 18.3K$

d) $A_v = \dfrac{-h_{fe} R_L}{h_{ie} + \Delta^h R_L} = \dfrac{-40 \times 2K}{1.5K + 48 \times 10^{-3} \times 2K} \doteq 50$

   where $\Delta^h = h_{ie} h_{oe} - h_{fe} h_{re} = 1.5K \times 40\mu v - 40 \times 3 \times 10^{-4} = 48 \times 10^{-3}$

e) $R_L = \left(\dfrac{h_{ie}}{h_{oe} \Delta^h}\right)^{1/2}$ optimum $= \left(\dfrac{1.5K}{40\mu v \times 48 \times 10^{-3}}\right)^{1/2} = 28K\Omega$

Solution 8:

$$f(t) = \begin{cases} \dfrac{10}{\pi} t & 0 < t < \pi \\ 0 & \pi < t < 2\pi \end{cases}$$

1. No symmetry

2. $a_0 = \dfrac{1}{2\pi} \int_0^{2\pi} f(t)\, dt = \dfrac{1}{2\pi} \int_0^{\pi} \dfrac{10}{\pi} t\, dt = \dfrac{5}{\pi^2} \dfrac{t^2}{2}\Big|_0^{\pi} = \dfrac{5\pi^2}{\pi^2 \cdot 2} = 5/2$

   $a_0 = 5/2$

3. $a_n = \dfrac{1}{\pi} \int_0^{2\pi} f(t) \cos n t\, dt = \dfrac{1}{\pi} \int_0^{\pi} \dfrac{10}{\pi} t \cos n t\, dt$

   $\dfrac{10}{\pi^2}\left[\dfrac{t}{n} \sin n t \Big|_0^{\pi} - \int_0^{\pi} \dfrac{\sin n t}{n}\, dt \right]$

   $\qquad u = t,\ dv = \cos n t\, dt$
   $\qquad du = dt \qquad v = \dfrac{\sin n t}{n}$

   $= \dfrac{10}{\pi^2} \dfrac{\cos n t}{n^2}\Big|_0^{\pi} = \dfrac{10}{n^2 \pi^2}(\cos n \pi - 1)$

   $a_n = \dfrac{10}{n^2 \pi^2}(\cos n \pi - 1) \qquad a_n = \begin{cases} \dfrac{-20}{n^2 \pi^2} & n \text{ odd} \\ 0 & n \text{ even} \end{cases}$

4. $b_n = \dfrac{1}{\pi} \int_0^{2\pi} f(t) \sin n t\, dt = \dfrac{1}{\pi} \int_0^{\pi} \dfrac{10}{\pi} t \sin n t\, dt$

   $\dfrac{10}{\pi^2}\left[-\dfrac{t}{n} \cos n t \Big|_0^{\pi} - \int_0^{\pi} -\dfrac{\cos n t}{n}\, dt \right]$

   $\qquad u = t \qquad dv = \sin n t\, dt$
   $\qquad du = dt \qquad v = -\dfrac{\cos n t}{n}$

Solution 8 cont'd

$$= \frac{10}{\pi^2} - \left[\frac{\pi}{n} \cos n\pi - 0 + \frac{\sin nt}{n^2}\Big|_0^\pi\right] = \frac{-10}{n\pi} \cos n\pi$$

$$b_n = -\frac{10}{n\pi} \cos n\pi \qquad b_n = \begin{cases} \frac{10}{n\pi} & n \text{ odd} \\ \frac{-10}{n\pi} & n \text{ even} \end{cases}$$

5. $f(t) = 5/2 - \frac{20}{\pi^2} \cos t - \frac{20}{9\pi^2} \cos 3t - \frac{20}{25\pi^2} \cos 5t \ldots$

$\qquad + \frac{10}{\pi} \sin t - \frac{10}{2\pi} \sin 2t + \frac{10}{3\pi} \sin 3t - \frac{10}{4\pi} \sin 4t + \ldots$

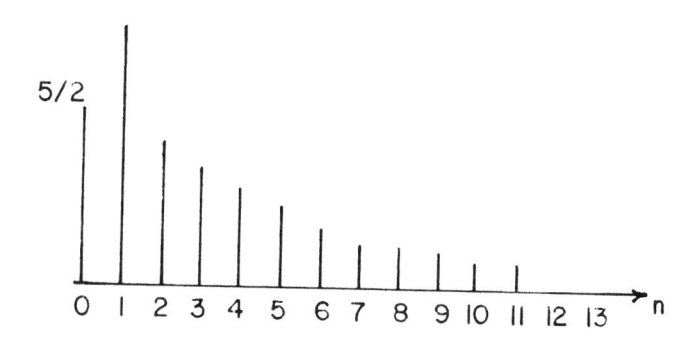

Solution 9:

$$I_1(3+j4) = I_2(10) \qquad (I_1)^2 \cdot 3 + (I_2)^2 10 = 1100$$

Mag only $\quad I_1(5) = I_2(10) \qquad (2I_2)^2 3 + I_2^2 10 = 1100$

$$I_1 = 2I_2 \qquad 12 I_2^2 + 10 I_2^2 = 1100$$

$$I_2^2 = \frac{1100}{22} = 50$$

$$I_2 = 7.07\underline{/0°}$$

Solution 9 cont'd

$5I_1 = 7.07(10)$

$I_1 = \dfrac{7.07}{5} = 14.14$

$I_1 = 14.14 \:\underline{/53.10°} = 8.49 + j11.30$

$A = I_1 + I_2 = 8.49 + j11.30 + 7.07 = 15.56 + j11.30$

$\qquad\qquad I_T = 19.23 \:\underline{/36°} \qquad\qquad 19.23 \:\underline{/36°}$

$P_1 = I_1^2 R_1 = (14.14)^2(3) = 600W \qquad P_1 = 600$ watts

$P_2 = I_2^2 R_2 = (7.07)^2(10) = 500 \qquad P_2 = 500$ watts

AFTERNOON SECTION

You will have four hours in which to work this test. Your score will be directly proportional to the number of problems you solve correctly through four (4). Each correct solution counts ten points. The maximum possible score for this part of the examination is 40 points. Partial credit for partially correct solutions will be given.

Work four of the problems according to instructions. Do not submit solutions or partial solutions for more than four problems. Indicate the problems which you have solved.

You may work only one engineering economy problem. When you have completed this portion of the P&P examination in the required time limit, you should check your solutions with the answers at the end of this examination. For additional practice you are encouraged to work the other problems under the time limit.

Since you want the maximum points available, you should remember that the examiner who assigns these points must make his judgment based only on what has been written down during the examination. It is important for you to be reasonably neat in your work and write down any assumption that you consider necessary to allow you to solve the problem properly and to provide sufficient rationale so that the examiner can judge your reasoning. Assumptions should follow the logic and requirements of the problem.

You are advised to use your time effectively.

1. A company proposes to install printing and marking equipment having a first cost of $48,000 and operating disbursements of $22,000 a year. The economic life is estimated to be 10 years with a salvage value of $5,400 at that date.

The company is considering alternative equipment for this job, costing $38,000 with operating disbursements of $25,000 a year. The economic life is also estimated to be 10 years with a salvage of $4,275 at that date.

The method of depreciation for tax and profit computation is sum-of-the-year's digits, and the company is permitted to use the same tax lives as the economic lives listed below as well as the same salvage values. The company's tax rate is 50%.

Using the method in which you compute the tax savings, find the rate of return on the extra investment.

2.
(a) Sketch the $V_{out} - V_{in}$ hysteresis loop for the following circuit. Determine V_{inL} and V_{inH}, the low and high switching points, in terms of V_{oH}, V_{oL}, V_R, R_1 and R_2.

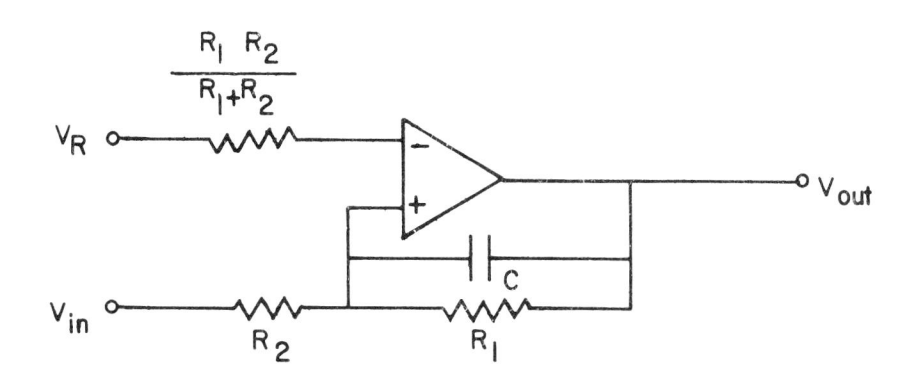

2 cont.
(b) Design the network such that the hysteresis loop is as shown below. (Assume on op amp output which comes within IV of the supply voltages.)

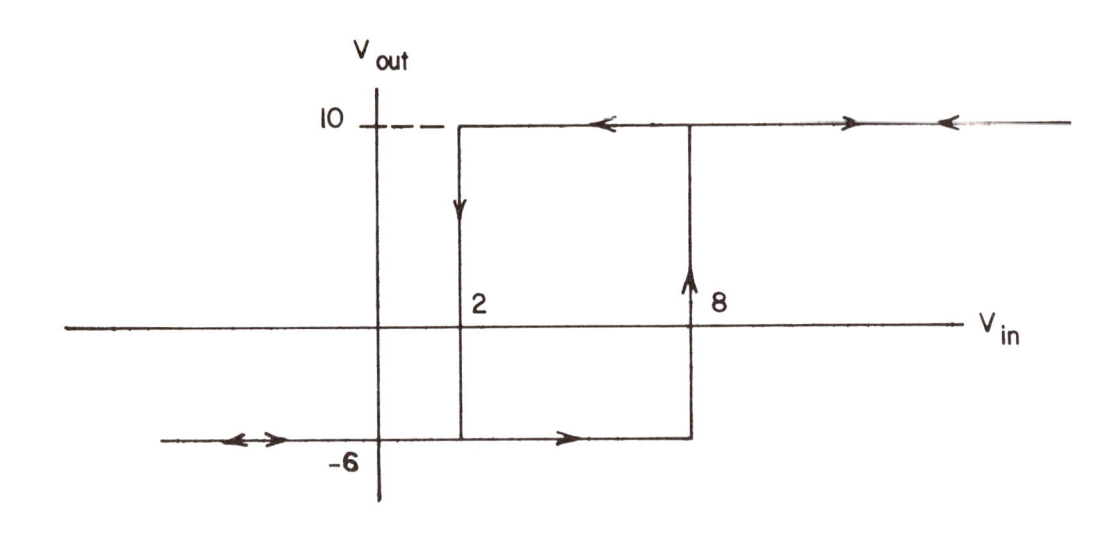

3. A dynamometer brake test on a fully loaded four-pole, 115-v single-phase motor running at rated speed revealed the following data: power drawn by motor, 150 w; input current, 2.0 amp; speed, 1750 rpm; length of brake arm, 12 in; dynamometer scale reading, 6 oz at rated load. When operated at an over-load to determine maximum torque, the following data was obtained: power drawn by motor, 550 w; input current, 10.0 amp; speed, 1400 rpm; dynamometer scale reading, 26.5 oz at maximum load. Calcuate

    a) Efficiency at rated load

    b) Power factor at rated load

    c) Motor horsepower and rated torque

    d) Maximum and starting torque

    e) Efficiency and power factor at maximum torque

    f) Rated torque from maximum torque

4. At high radio frequencies, coaxial transmission lines are customary. At these frequencies, the attenuation constant is usually small compared with the phase constant and the characteristic impedance is almost the same as that for a lossless line.

Consider a common flexible coaxial cable, designated RG8U, at a frequency of 100 Mc. At this frequency,

$$R = 0.79 \text{ ohm/m}$$
$$L = 0.262 \text{ μh/m}$$
$$G = 1.93 \times 10^{-6} \text{ mho/m}$$
$$C = 96.8 \text{ μμf/m}$$

Find the propagation constant, attenuation constant, phase constant, wavelength, velocity of propagation and characteristic impedance.

5.
a) Determine the transfer function of a system which has a gain factor of 3 and the pole-zero map as shown.

b) Evaluate the step response of the system.

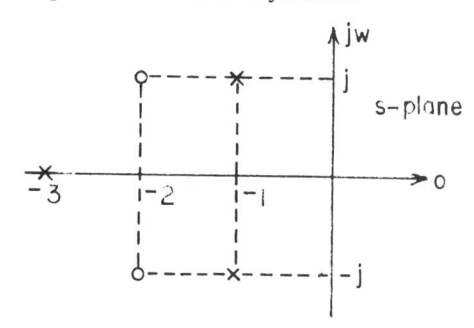

[s-plane pole-zero map: poles (×) at -3, $-2+j$, $-2-j$; zeros (○) at $-3+j$, $-3-j$]

6. The owner of an industrial plant has requested an illumination design for the plant represented by the figure. You are assigned low bay area #1 excluding the office. The ceiling is unobstructed with a ceiling height of 14 ft. A 4 ft. industrial type luminaire utilizing either two of three 40-watt 48T12 cool white fluorescent lamps or a corresponding 8 ft. unit has been chosen for the area. Luminaire suspension is to be 2.5 feet. The illumination design level is 50 ft. candle maintained. The coefficient of utilization for the luminaire and area is 0.53, and the maintenance factor is 0.65. The spacing to mounting height is not to exceed 1.0. Assume no contribution from other parts of plant.

a) Calcuate the number of equivalent 4-ft. luminaires required. (one 8-ft. luminaire is equivalent to two 4-ft. luminaires.) Be sure to specify either 2-lamp or 3-lamp luminaires.

6 cont'd

b) Give a layout sketch will will indicate luminaire location, orientation, arrangement, spacing, and 8-ft. or 4-ft. length. Dimension spacing of luminaire-to-wall, luminaire-to-column, and luminaire spacing as applicable to a meaningful layout.

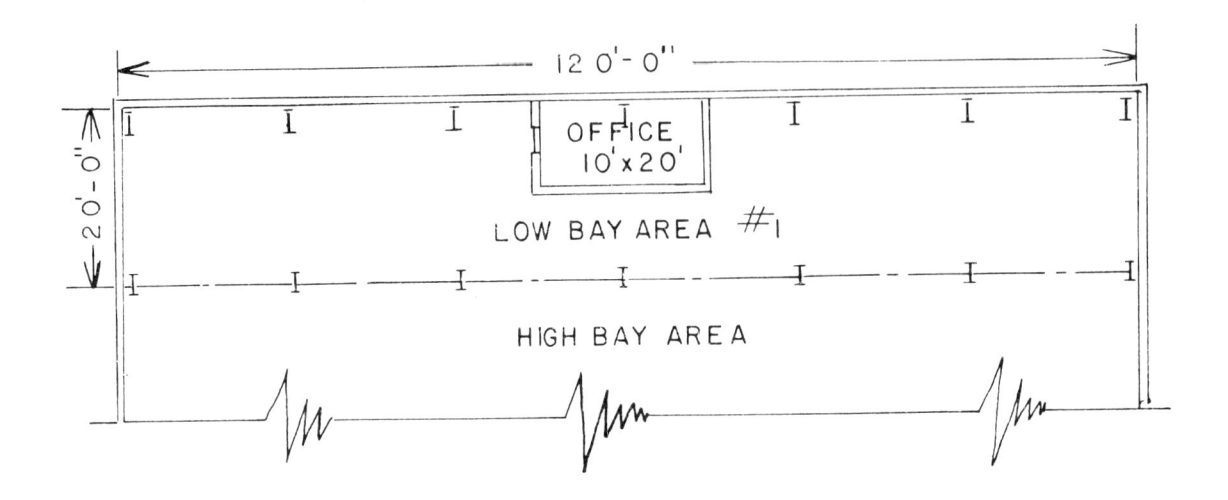

7. Determine the approximate circuit component values and V_{CC} supply for a 2-MH$_Z$, NPN transistor, Hartley oscillator, similar to the one shown when the following conditions and circuit values are identified:

a) $A_v \beta = 3, S = 6, R_E = 1K\Omega$

b) The combined choke and coil resistance is 10Ω

c) $h_{ie} = 1K\Omega$, $h_{fe} = 60$, $h_{oe} = 10^{-6}$ mho, $h^e = 0.02$

d) $V_{CE} = 05V$, $I_C = -3mA$, $I_B = 50\mu A$, $V_{BE} = -0.2V$

e) $L_1 = 5\mu H$, I_{CO} is negligible

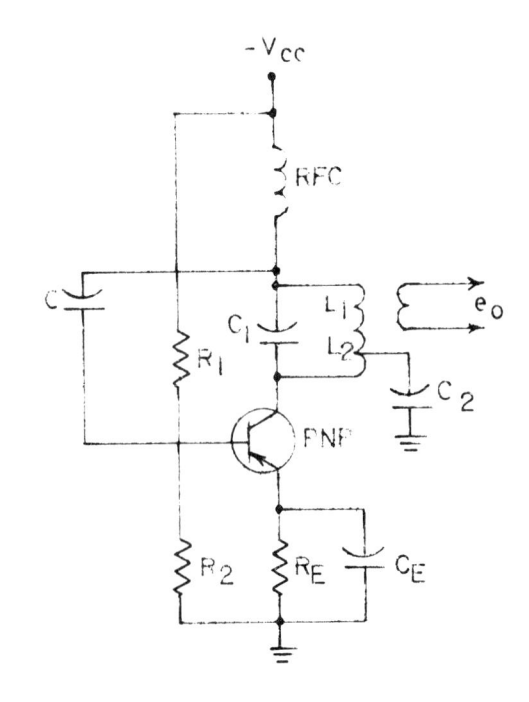

8.
a) Matching in the circuit shown is done with a unity-coupled coil instead of an ideal transformer. At very low frequencies the power is shunted by the transformer and is not delivered to the load. The matching is adjusted to give a maximum power in the load for high frequencies, where the shunting effect of the transformer can be neglected. Find the turns-ratio. The primary inductance remains at 1 h as the secondary turns are varied. Give also L_2 and M.

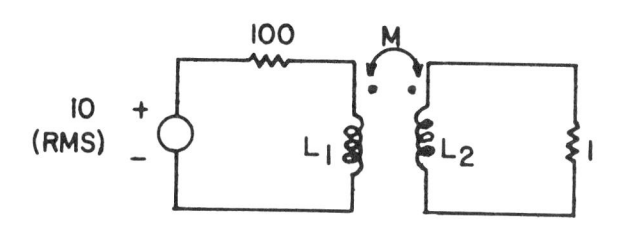

b) In the circuit shown find the value of the capacitive reactance which will give unity power factor at the terminals of the generator, although connected directly across the load.

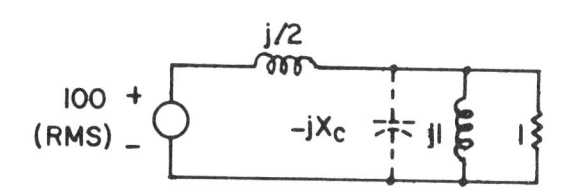

9. A three-phase, three-wire, 416 volt, CBA system has a wye-connected load with $Z_A = 6\underline{|0°}$, $Z_B = 3\underline{|30°}$, and $Z_C = 5\underline{|45°}$. Obtain the line currents and the phasor voltage across each impedance.

Solution 1:

	First Year		Subsequent Years	
	A	B	A	B
Op. Disb. BT	22,000	25,000		
Depr.(P-L) $\frac{2 \times 10}{10 \times 11}$	7,745	6,131.60	ec.dep. 774.50	613.16
Total Expense	29,745	31,131.60	extra tax 387.25/yr.	306.58/yr
Tax Saving	14,872.50	15,565.80		
Op. Disb. AT	7,127.50	9,434.20		

$(48,000-5,400)(crf) + 5,400i + 7,127.50 + 387.25 \text{ (asf)}$
$10 10$

$\overset{\wedge}{=} (38,000-4,275)(crf) + 4,275i + 9,434.20 + 306.58(asf)$
$10 10$

Whence i = 16.3%

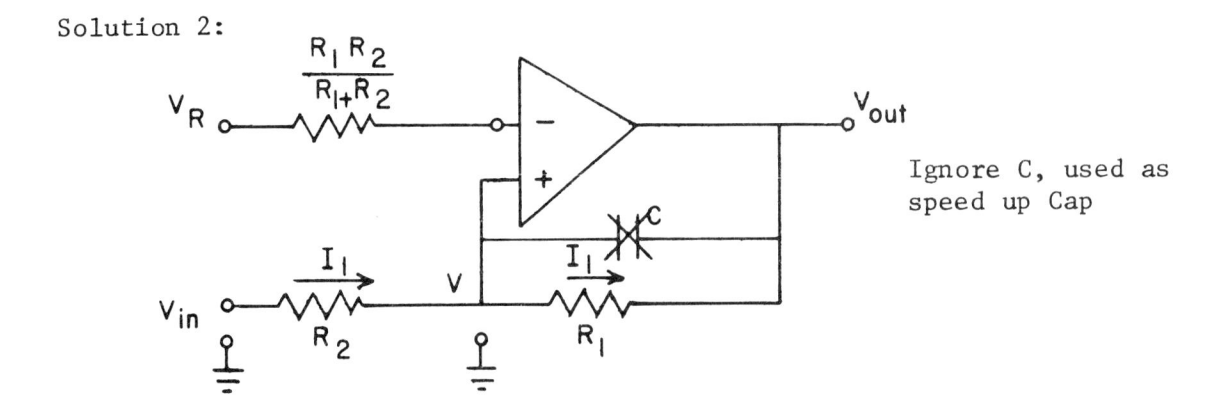

Solution 2:

Ignore C, used as speed up Cap

A) Sketch the $V_{out} - V_{in}$ hysteresis loop. Determine V_{inL} and V_{inH}, (the low & high switching points) in terms of V_{oH}, V_{oL}, V_R, R_1 and R_2.

Solution 2 cont'd

Assume $V_{in} < V_R$ and increasing
when $V_{in} = V_{inH}$, the voltage $V = V_R$

$$\frac{V_{oL} - V_R}{R_1} = \frac{V_R - V_{inH}}{R_2}$$

$$V_{inH} = V_R - \frac{R_2}{R_1}(V_{oL} - V_R)$$

and similarly when $V_{in} = V_{inL}$, $V = V_R$

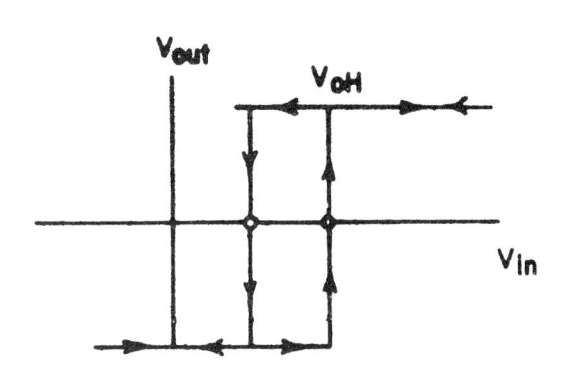

$$V_R = 4.18V$$

$$\frac{V_{oH} - V_R}{R_1} = \frac{V_R - V_{inL}}{R_2} \qquad \frac{R_2}{R_1} = \frac{V_R - V_{inL}}{V_{oH} - V_R} = \frac{4.18 - 2}{10 - 4.18} = \frac{2.18}{5.82} = .372$$

Solving Simul. $V_{inL} = V_R - (V_{oH} - V_R)\frac{R_2}{R_1} \qquad \frac{R_2}{R_1} = .372$

$$\frac{V_{oL} - V_R}{V_{oH} - V_R} = \frac{V_R - V_{inH}}{V_R - V_{inL}}$$

$$\frac{-6 - V_R}{10 - V_R} = \frac{V_R - 8}{V_R - 2} \qquad 22V_R = 92$$

$$V_R = \frac{92}{22} = \underline{4.18V}$$

b) Design (a) to match the Hyst-Loop below:

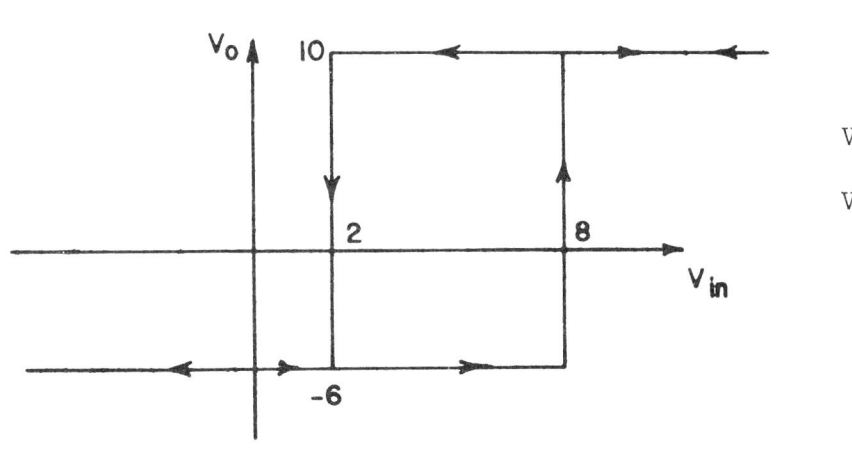

$V_{oH} = 10$

$V_{oL} = -6$

$V_{inH} = 8$

$V_{inL} = 2$

202 / SECOND SAMPLE P & P/EE EXAMINATION

Solution 2 cont'd

$$\frac{V_{inH}-V_R}{V_{inL}-V_R} = \frac{V_R-V_{oL}}{V_R-V_{oH}} \qquad \frac{8-V_R}{-6-V_R} = \frac{V_R-(-6)}{V_R-8}$$

$$-(V_R-8)^2 = -(V_R+6)^2$$

$V_R = 4.18V$

$\dfrac{R_2}{R_1} = .372$

Solution 3:

a) Watts output $= \dfrac{TS}{5252} \times 746\dfrac{w}{hp} = \dfrac{6\ oz}{\dfrac{16oz}{1b}} \times \dfrac{1\ ft \times 1750}{5252} \times 746 = 93.3$ w.

   Effy $= \dfrac{\text{Watts Out}}{\text{Watts In}} = \dfrac{93.3}{150} = -/622$ at rated load

b) $\cos\theta = \dfrac{W}{EI} = \dfrac{150\ w}{115\times 2} = 0.5$

c) hp $= \dfrac{TS}{5252} = \dfrac{6 \times 1750}{16 \times 5252} = \dfrac{1}{8}$ hp

   $T_{rated} = 6\ oz \times 12\ in = 72\ oz\ in.$

d) $T_{max} = 26.5\ oz \times 12\ in. = 293\ oz\ in.$, $s = \dfrac{1800-1400}{1800} = 0.222$

   $T_s = T_b \dfrac{2}{(\dfrac{0.222}{1.0} + \dfrac{1.0}{0.222})} = 293\ oz\ in.\ \dfrac{2}{4.722} = 124.2\ oz\ in.$

e) Effy $= \dfrac{\text{Watts Out}}{\text{Watts In}} = \dfrac{TS}{5252} \times \dfrac{746}{550_N} = \dfrac{293\ oz\ in.}{16 \times 12} \times \dfrac{1400}{5252} \times \dfrac{746}{550} = 0.553$

   $\cos\theta = \dfrac{W}{EI} = \dfrac{550}{115\times 10} = 0.478$

f) $T_r = T_b \dfrac{2}{(\dfrac{s_b}{s} + \dfrac{s}{s_b})} = 293 \dfrac{2}{(\dfrac{0.222}{0.0278} + \dfrac{0.0278}{0.222})} = 72.2\ oz\ in.$

   rated slip, $\underline{s} = \dfrac{1800-1750}{1800} = 0.0278$

Solution 4:

$$\gamma = \sqrt{(R+jwL)(G+jwc)} = \alpha+j\beta$$

$$\gamma = \sqrt{(0.79+j2\pi \times 10^8 \times 0.262 \times 10^{-6}) \times (1.93 \times 10^{-6}+j2\pi \times 10^8 \times 0.968 \times 10^{-10})}$$

$$= \sqrt{(0.79+j164)(1.93 \times 10^{-6}+j6.08 \times 10^{-2})}$$

$$= j\sqrt{(1.64 \times 6.08)(1-j\frac{0.79}{1.64} \times 10^{-2})(1-j\frac{1.93}{6.08} \times 10^{-4})}$$

$$= j3.15\sqrt{(1-j4.8 \times 10^{-3})(1-j3.18 \times 10^{-5})}$$

$$\simeq j3.15(1-j4.8 \times 10^{-3})^{\frac{1}{2}}$$

The binomial theorem states
$$(1+x)^{\frac{1}{2}} = 1+\tfrac{1}{2}x+ \dots$$

So

$$\gamma \simeq j3.15(1-j2.4 \times 10^{-3}) = 7.6 \times 10^{-3}+j3.15$$

$$\alpha = 7.6 \times 10^{-3} \text{ nepers/m} = 8.686 \times 7.6 \times 10^{-3} = 6.62 \times 10^{-2} \text{ db/m}$$

$$\beta = 3.15 \text{ rad/m}$$

$$\lambda = \frac{2\pi}{\beta} = \frac{2\pi}{3.15} = 0.99 \text{ m}$$

$$v_P = W/\beta = \frac{2\pi \times 10^8}{3.15} = 1.99 \text{ m/sec}$$

$$Z_o = \sqrt{\frac{R+jwL}{G+jwC}} = \sqrt{\frac{0.79+j164}{1.93 \times 10^{-6}+j6.08 \times 10^{-2}}} \simeq \sqrt{\frac{164}{6.08 \times 10^{-2}}} = 52 \text{ ohms}$$

Solution 5:

a) The transfer function has zeros at $-2 \pm j$ and poles at -3 and at $-1 \pm j$. The transfer function is therefore

$$P(s) = \frac{3(s+2+j)(s+2-j)}{(s+3)(s+1+j)(s+1-j)}$$

Solution 5 cont'd

b) The Laplace transform of the system output is

$$Y(s) = P(s)X(s) = \frac{3(s+2-j)(s+2-j)}{s(s+3)(s+1+j)(s+1-j)}$$

Expanding $Y(s)$ into partial fractions yields

$$Y(s) = \frac{R_1}{s} + \frac{R_2}{s+3} + \frac{R_3}{s+1+j} + \frac{R_4}{s+1-j}$$

where $R_1 = \frac{3(2+j)(2-j)}{3(1+j)(1-j)} = \frac{5}{2}$

$R_2 = \frac{3(-1+j)(-1-j)}{-3(-2+j)(-2-j)} = \frac{-2}{5}$

$R_3 = \frac{3(1)(1-2j)}{(-1-j)(2-j)(-2j)} = \frac{-3}{20}(7+j)$

$R_4 = \frac{3(1+2j)(1)}{(2+j)(-1+j)(2j)} = \frac{-3}{20}(7-j)$

Evaluating the inverse Laplace transform,

$$y = \frac{5}{2} - \frac{2}{5}e^{-3t} - \frac{3\sqrt{2}}{4}e^{-t}\left[e^{-j(t+\theta)}+e^{j(t+\theta)}\right] = \frac{5}{2} - \frac{2}{5}e^{-3t} - \frac{3\sqrt{2}}{2}e^{-t}\cos(t+\theta)$$

where $\theta = -\tan^{-1}(1/7) = -8.13°$.

Solution 6:

Industrial plant - fluorescent luminaire:

No. of 3-lamp luminaires $= \frac{E \times A}{n \times \phi \times C.O.U. \times M.F.}$

$= \frac{50(20 \times 120)}{3 \times 2650 \times 0.53 \times 0.65} = 44$

No. of 8-ft, 6-lamp luminaires = 22

No. of 4-ft, 2-lamp luminaires = 66

No. of 8-ft, 4-lamp luminaires = 33

Solution 6 cont'd

Use four 8-ft. units per bay (24 units) less 2 units for office area for a total of 22 8-ft. units.

Preferred Layouts: Acceptable Layouts:

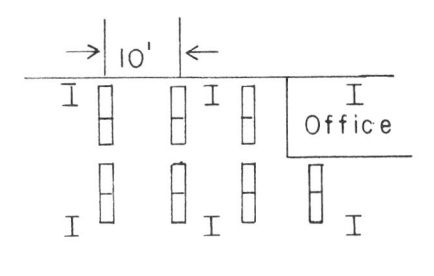

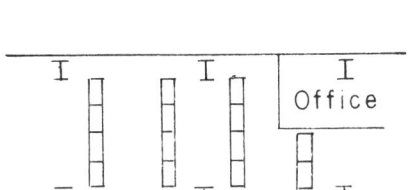

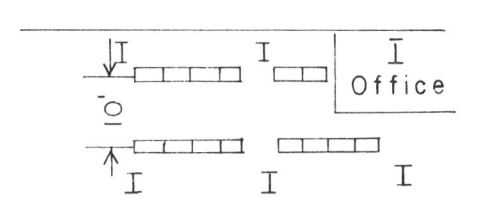

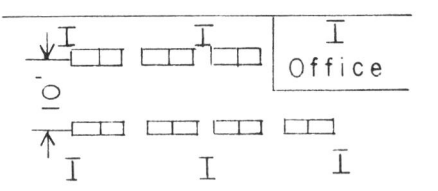

I_n - service $E = \dfrac{24}{22} \times 50 = 54.5$ ft.c.

Solution 7:

$R_2 \doteq (S-1)R_E = (6-1)1K = 5K\Omega$ $\qquad$ $V_{R_E} \doteq I_C R_E = 3\text{mA} \times 1K\Omega = 3V$

$V_{R_2} = V_{R_E} + V_{BE} = 3V + 0.2V = 3.2V$

$V_{CC} = V_{R_E} + V_{CE} + I_C R_{DC} = 3V + 5V + 3\text{mA} \times 10\Omega \qquad V_{CC} \doteq 8V$

$V_{R_1} = V_{CC} - V_{R_2} = 8V - 3.2V = 4.8V \qquad I_{R_2} = \dfrac{V_{R_2}}{R_2} = \dfrac{3.2V}{5K\Omega} = 0.64\text{mA}$

$I_{R_1} = I_{R_2} + I_B = 0.64\text{mA} + 50\mu A = 690\mu A \qquad R_1 = \dfrac{V_{R_1}}{I_{R_1}} = \dfrac{4.8V}{0.69\text{mA}} = 6.95K\Omega$

$\dfrac{h_{fe} L_1}{\Delta h^e L_2} = 3, \; L_2 = \dfrac{h_{fe} L_1}{3\Delta h^e} = \dfrac{60 \times 5\mu H}{3 \times 0.02} = 5\text{mH} \qquad L_T \doteq L_1 + L_2 = 5\mu H + 5\text{mH} \doteq 5\text{mh}$

Solution 7 cont'd

$$C_1 = \frac{1}{4\pi^2 f_o^2 L_T} = \frac{1}{4\pi^2 \times 4\times 10^{12} \times 5mH} = 1.25 pF$$

$$X_{C_E} = \frac{1}{20} R_E @ f_o = \frac{1}{20} 1K\Omega = 50\Omega \qquad C_E = \frac{1}{2\pi f_o X_{CE}} = \frac{0.159}{2\times 10^6 \times 50} = 1.59 nF$$

$$X_C = \frac{1}{20} h_{ie} @ f_o = \frac{1}{20} 1K\Omega = 50\Omega$$

$$\therefore C = 1.59 \text{ nF} \qquad X_{RFC} = 20 h_{ie} @ f_o = 20 \times 1K\Omega = 20 K\Omega$$

$$L_{RFC} = \frac{X_{RFC}}{2\pi f_o} = \frac{20 K\Omega}{2\pi \times 2 \times 10^6} = 1.6 mH \qquad X_{C_2} = \frac{1}{20} 1K\Omega = 50\Omega$$

$$\therefore C_2 = 1.59 nF$$

Solution 8:

a) $L_1 = 1h \qquad N = \sqrt{\dfrac{L_2}{L_2}}$

for power matching Z_2 reflected must be equal to 100

$$Z_2 = N^2 Z_1$$
$$N^2 = \frac{Z_2}{Z_1} = \frac{1}{100}$$
$$N = \frac{1}{10}$$

$$L_2 = N^2 L_1 = 10^2 (1) = 100$$

unity coupling = $K = 1$, $M = \sqrt{L_1 L_2} = \sqrt{1 \cdot 100} = 10$

Solution 8 cont'd

b)

Must make I_m part of Z to right of a-a' be equal $-\frac{1}{2}$ to cancel the j/2 to the left of a-a'

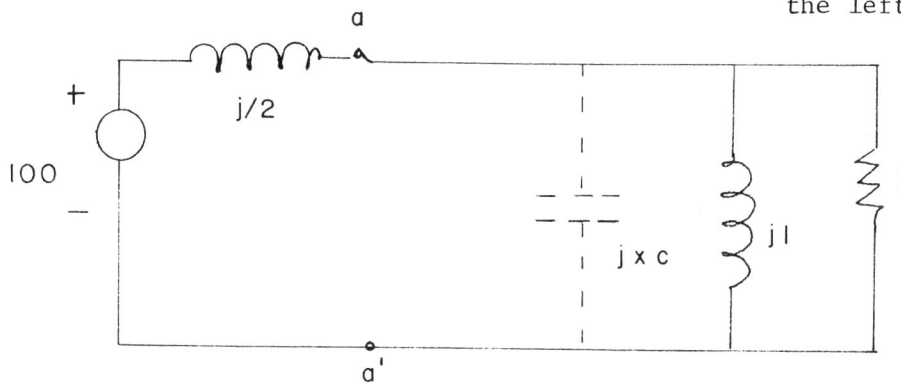

$$Y = \frac{1}{-jX_C} + \frac{1}{j1} + \frac{1}{1} = j\frac{1}{X_C} - j1 + 1$$

$$Y = 1 + j(\frac{1}{X_C} - 1)$$

$$Z = \frac{1}{1+j(\frac{1}{X_C} - 1)} \times \text{conjugate} = \frac{1-j(\frac{1}{X_C} - 1)}{1^2+(\frac{1}{X_C} - 1)^2}$$

Now $I_m Z = \dfrac{-(\frac{1}{X_C} - 1)}{1+(\frac{1}{X_C} - 1)^2} = -\frac{1}{2}$

$$\frac{2}{X_C} - 2 = 1 + \frac{1}{X_C^2} - \frac{2}{X_C} + 1$$

$$\frac{1}{X_C^2} - \frac{4}{X_C} + 4 = 0$$

$$(\frac{1}{X_C} - 2)^2 = 0$$

$$\frac{1}{X_C} = 2$$

$$X_C = \frac{1}{2}$$

Solution 9:

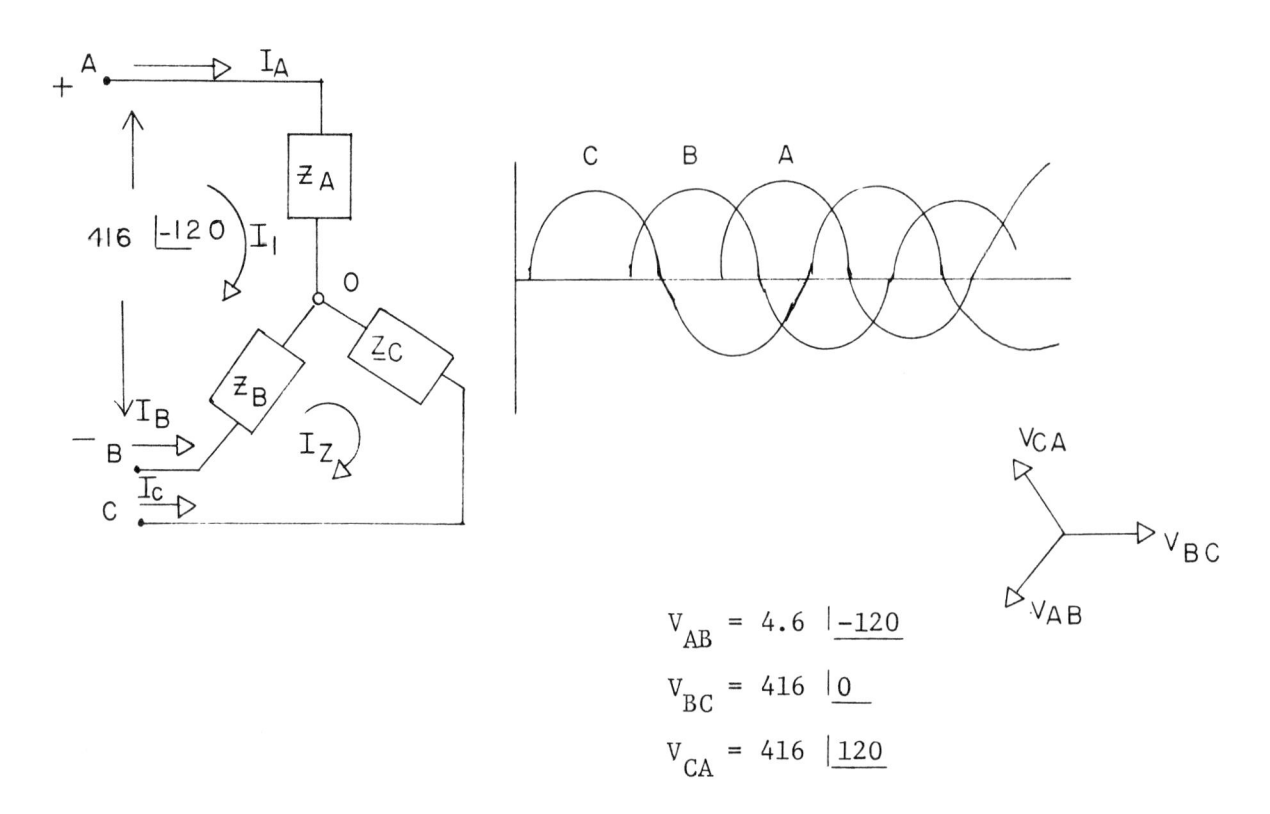

$V_{AB} = 4.6 \angle -120$
$V_{BC} = 416 \angle 0$
$V_{CA} = 416 \angle 120$

$Z_A = 6\angle 0 = 6+j0$, $Z_B = 6\angle 30 = 5.2+j3$, $Z_C = 5\angle 45 = 3.54+j3.54$
$V_{BA} = 416 \angle 120 = -208-j360.27$, $V_{CB} = 416\angle 0° = 416+j0$

Write Mesh E_Q

$I_1(Z_A+Z_B) - I_2 Z_B = V_{BA}$

$-I_1(Z_B)+I_2(Z_B+Z_C) = V_{CB}$

$I_1(6+5.2+j3)-I_2(6\angle 30) = 416\angle -120$

$-I_1(6\angle 30)+I_2(5.2+j3+3.54+j3.54) = 416$
$\qquad\qquad\qquad\qquad\qquad 8.74+j6.54$

$I_1(11.59\angle 15)-I_2(6\angle 30) = 416\angle -120$

$-I_1(6\angle 30)+10.92\angle 36.81 = 416$

Solution 9 cont'd

$$I_1 = \frac{\begin{vmatrix} 416\underline{|-120} & -6\underline{|30} \\ 416 & 10.92\underline{|36.81} \end{vmatrix}}{\begin{vmatrix} 11.59\underline{|15} & -6\underline{|30} \\ -6\underline{|30} & 10.92\underline{|36.81} \end{vmatrix}} = \frac{+4542.72\underline{|-83}+2496\underline{|30}}{126.56\underline{|51.81} - 36\underline{|60}}$$

$$= \frac{553.62-j4508.86+2161.60+j1248}{78.29+j99.47-18-j31.18}$$

$$I_1 = \frac{2715.22-j3260.86}{60.29+j68.29} = \frac{4243.3\underline{|-50.22}}{91.1\underline{|48.56}} = 46.58\underline{|-98.78}$$

$$I_1 = 46.58\underline{|-98.78}$$

$$I_2 = \frac{\begin{vmatrix} 11.59\underline{|15} & 416\underline{|-120} \\ -6\underline{|30} & 416 \end{vmatrix}}{\Delta} = \frac{416(11.59\underline{|15}+6\underline{|-90})}{\Delta}$$

$$= \frac{416}{\Delta}(11.3+j3-j6)$$

$$= \frac{416}{\Delta}(11.2-j3) = \frac{416(11.59\underline{|-15}}{91.1\underline{|48.56}}$$

$$I_2 = 52.92\underline{|-63.56}$$

$I_A = I_1 = 46.58\underline{|-98.78}$

$I_C = -I_2 = 52.92\underline{|117.44}$

$I_B = I_2 - I_1 = 52.92\underline{|-63.53} - 46.58\underline{|-98.78}$

$\quad\quad = 23.56-j47.38+7.11+j46.03$

$\quad\quad = 30.67-j1.35$

$\quad\quad I_B = 30.71\underline{|-2.52}$

$V_{AO} = I_A Z_A = 46.58\underline{|-98.78} \times 6\underline{|0} = 279.48\underline{|-98.78} = 279.48\underline{|261.22}$

$V_{BO} = I_B Z_B = 30.71\underline{|-2.52} \times 6\underline{|30} = 184.26\underline{|27.48}$

$V_{CO} = I_C Z_C = 52.92\underline{|117.44} \times 5\underline{|45} = 264.60\underline{|162.44}$

Part IV

Third Sample Principles and Practice of Electrical Engineering Examination

MORNING SECTION

You will have four hours in which to work this test. Your score will be directly proportional to the number of problems you solve correctly through four (4). Each correct solution counts ten points. The maximum possible score for this part of the examination is 40 points. Partial credit for partially correct solutions will be given.

Work four of the problem according to instructions. Do not submit solutions or partial solutions for more than four problems. Indicate the problems which you have solved.

You may work only one engineering economy problem. When you have completed this portion of the P&P examination in the required time limit, you should check your solutions with the answers at the end of this examination. For additional practice you are encouraged to work the other problems under the time limit.

Since you want the maximum points available, you should remember that the examiner who assigns these points must make his judgment based only on what has been written down during the examination. It is important for you to be reasonably neat in your work and write down any assumption that you consider necessary to allow you to solve the problem properly and to provide sufficient rationale so that the examiner can judge your reasoning. Assumptions should follow the logic and requirements of the problem.

You are advised to use your time effectively.

1. Machine A costs $10,000 with operating disbursements of $4,950 a year for 8 years with $2,500 salvage at that time. Machine B costs $8,000 with operating disbursements of $5,500 a year with $2,000 salvage at the end of 8 years. The minimum required rate of return is 15%. (a) By the annual-cost method of analysis, which machine should be selected? (b) If management decrees the adoption of 4-year economic lives with zero salvage for all equipment, which machine must be selected? Describe the loss to the company, if any. (c) Suppose the economic life of the machine slected in part (a) does turn out to be 4 years, wnat will be the loss to the company, if any?

2. With switch S open, the following system is to produce the square and triangular wave shown. Determine values for R and C. For the values of R and C selected, sketch the outputs when the switch S is closed. (The diode is ideal.)

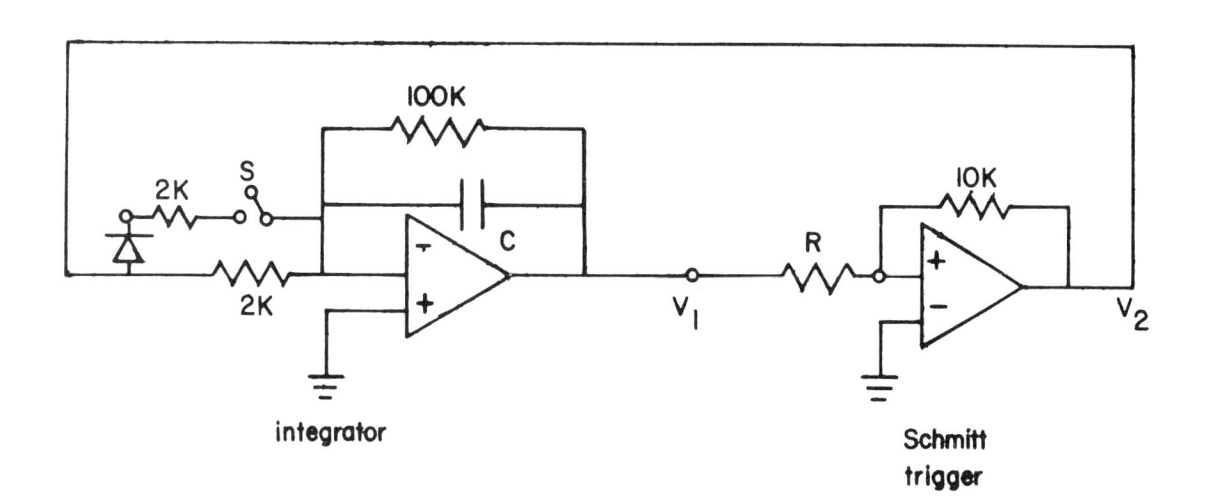

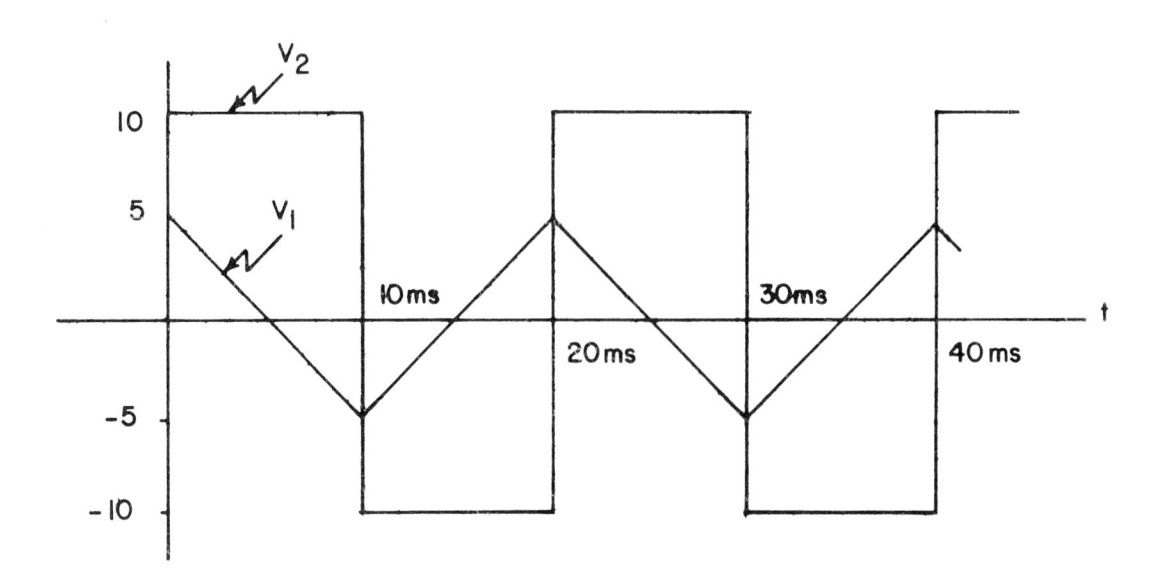

3. The results of three short-circuit tests on a 7,960:2,400:600-volt 60-cps single-phase transformer are as follows:

Test	Winding excited	Winding short-circuited	Applied voltage, volts	Current in excited winding, amp
1	1	2	252	62.7
2	1	3	770	62.7
3	2	3	217	208

Resistances may be neglected. The rating of the 7,960-volt primary winding is 1,000 kva, of the 2,400-volt secondary is 500 kva, and of the 600-volt tertiary is 500 kva.

a) Compute the per-unit values of the equivalent-circuit impedances of this transformer on a 1,000-kva rated-voltage base.

b) Three of these transformers are used in a 3,000-kva Y-Δ-Δ 3-phase bank to supply 2,400-volt and 600-volt auxiliary power circuits in a generating station. The Y-connected primaries are connected to the 13,800-volt main bus. Compute the per-unit values of the steady-state short-circuit currents and of the voltage at the terminals of the tertiary windings with 13,800 volts maintained at the primary line terminals. Use a 3,000-kva 3-phase rated-voltage base.

4.

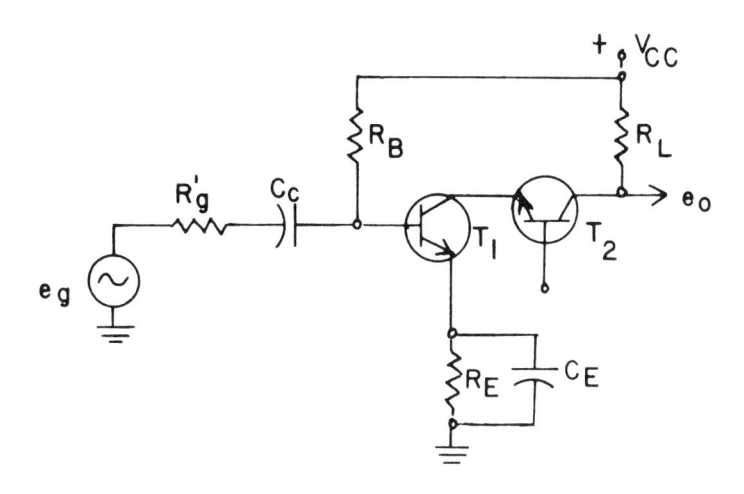

The transistor cascode amplifier has the following features:

For T_1:
$$h_{fe} = 100 \qquad h_{ie} = 1.5 \text{ K}\Omega \quad (\doteq r_{in_1})$$
$$C_{b'e} = 8 \text{ pF} \qquad C_{b'c} = 2 \text{ pF}, \quad C_W \text{ negligible}$$

For T_2:
$$r_{in_2} = 40\Omega \qquad R_L = 2\text{K}\Omega$$

Find:
    a) Overall voltage gain

    b) The upper breakpoint-frequency

    c) The gain-bandwidth product

5. Given the following transfer functions
$$G = \frac{K(s+5)}{s(s+1)(s+10)} = \frac{K(s+5)}{s^3 + 11 s^2 + 10 s}$$

    a) Find $\dfrac{C}{R} = \dfrac{G}{1+G}$

    b) Make a flow diagram for (a)

    d) Write a state space form from (b) for the system.

6. A multifamily dwelling has 20 apartments and a total floor area of 16,000 sq. ft. The general lighting load per apartment is 2400 watts. Each apartment is equipped with an electric range with a rating not exceeding 12kw. The apartments have equal floor areas. The building is supplied with power from a single-phase 11/230 volt, three-wire system. Use National Electrical Code for appliance loads, diversity factors, etc.

(a) Determine the maximum-demand present loads for each apartment and specify the size and type of insulated copper conductors enclosed in conduit for the main feeder to each apartment.

(b) Determine the maximum-demand for feeder supplying the building and specify the size and type insulated cooper conductors for the main feeder to the building. (To be enclosed in conduit).

7. The N-channel junction FET in the PA circuit has the drain family characteristic shown. This devise has a maximum drain dissipation ($P_{D_{mas}}$) of 160 mW with a case temperature of $50°C$.

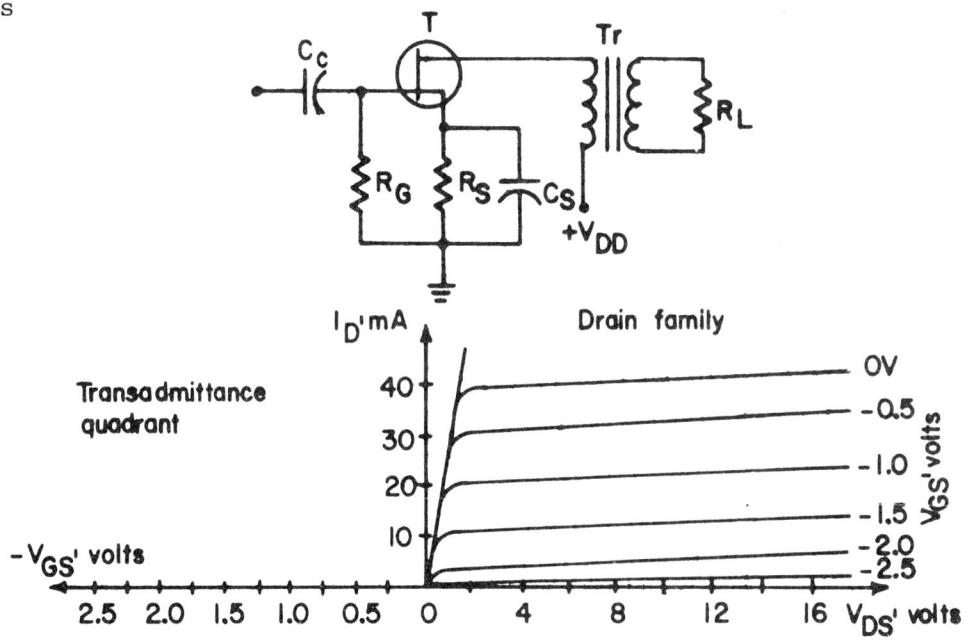

a) V_{DS} and R'_L for maximum signal excursion
b) Turns ratio n to transform $R_L = 10\Omega$ into R'_L
c) Power output graphically, when v_{gs} = 1V peak to peak.
d) Insertion power and power output.
e) Percentage of second-harmonic distortion for this signal in (c).
f) Power output.

8.

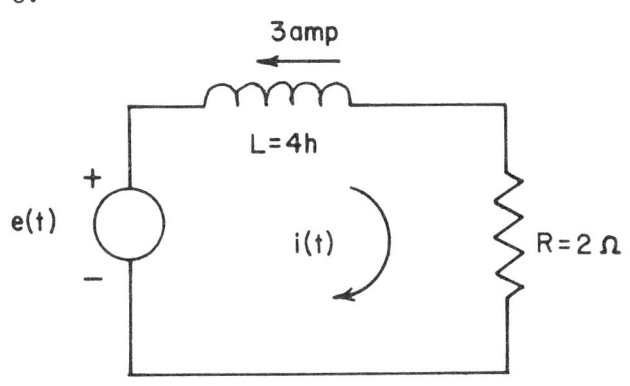

[Circuit: e(t) source with + on top, connected to loop containing L=4h inductor (with 3 amp arrow pointing left) on top branch and R=2Ω resistor on right branch; mesh current i(t) shown]

Given that e(t) = 4t for t ≥ 0 find:

a) i(t)

b) t when i(t) = 0

9.
a) A generator with a voltage of 120 v, rms, is feeding a load through a transmission line represented by the impedance 1+j. The power factor of the load is known to be 0.8 and the voltage across the load is 100 volts, rms. Find the current and the power factor at the generator.

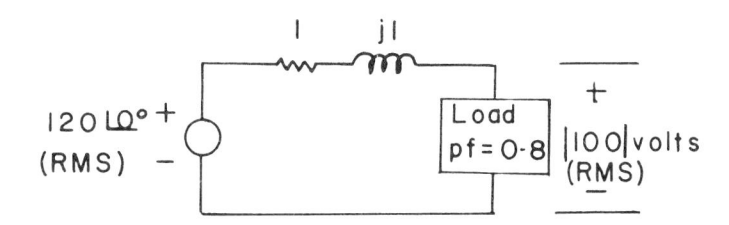

[Circuit: 120∠0° (RMS) source in series with 1 ohm resistor and j1 inductor feeding Load with pf=0.8, |100| volts (RMS) across load]

b) The circuit shown represents a generator feeding a load through a transmission line. The power factor of the load is 0.8 and the power factor at the generator is 0.5, without the capacitance. Find the capacitive reactance which then added across the generator will bring the power factor at the generator to unity.

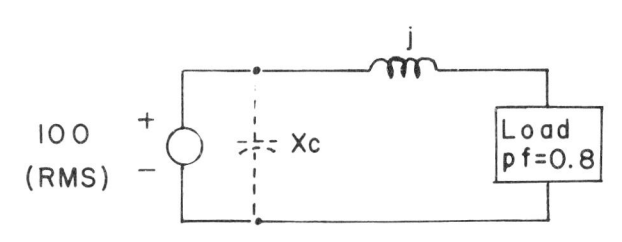

[Circuit: 100 (RMS) source with Xc capacitor in parallel across it, then in series with j inductor feeding Load with pf=0.8]

218 / THIRD SAMPLE P & P/EE EXAMINATION

Solution 1:

a) $AC_{A-B} = 1,500 \; (crf)^{.22285}_{.15-8} + 500(.15) - 550$

$AC_{A-B} = \$-140.72$ Select A

b) $AC_{A-B} = 2,000 \; (crf)^{.35026}_{.15-4} - 550$

$AC_{A-B} = \$150.52$ Select B

The loss is $140.72 a year plus 15% return on the $2,000 investment for 8 years.

c) The loss is $150.52 a year, the amount the investment fails to earn, minus 15% return on the $2,000 investment for 4 years.

Solution 2:

a) Switch open

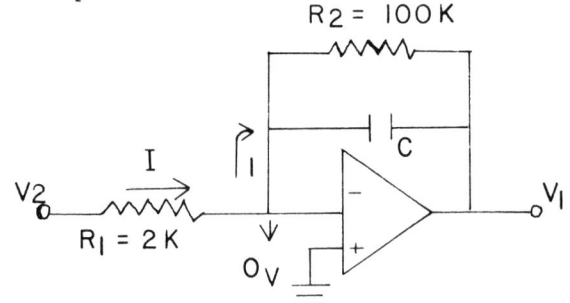

$Z_{in} = \infty$

$Z_{out} = 0$

$v_1 = Av_2 \;\; (A>0)$

$I = \dfrac{0-v_2}{R_1}$ $I = C\dfrac{d(v_i - 0)}{dt}$

Assuming the current through R_2 is negligible. Or that charge in cap is not changed over a full cycle.

$\dfrac{-v_2}{R_1} = C\dfrac{dv_i}{dt}$

$C = \dfrac{-v_2}{R_1 \, dv_1/dt} = -\dfrac{10}{2K(-10/10ms)} = 5\mu f$ where $\dfrac{dv_1}{dt} = -1v/ms$

$C = 5\mu f$

Solution 2 cont'd

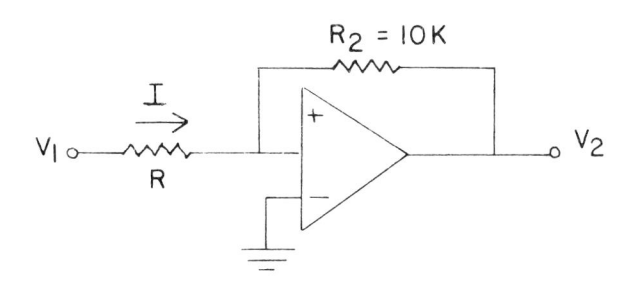

$$I = \frac{v_1}{R} = \frac{v_2}{R_2} \qquad R = \frac{v_1}{v_2} R_2 \qquad R = \frac{5}{10} \cdot 10K = 5K$$

R = 5K

b) Switch closed. Assume amplifiers maintain their output levels for v_2 positive $R_1 = 2K || 2K = 1K$

$R_1 = 1K$

$$\frac{dv_1}{dt} = \frac{-v_2}{R_1 C} = \frac{-10}{1K \times 5\mu f} = \frac{-2V}{ms}$$

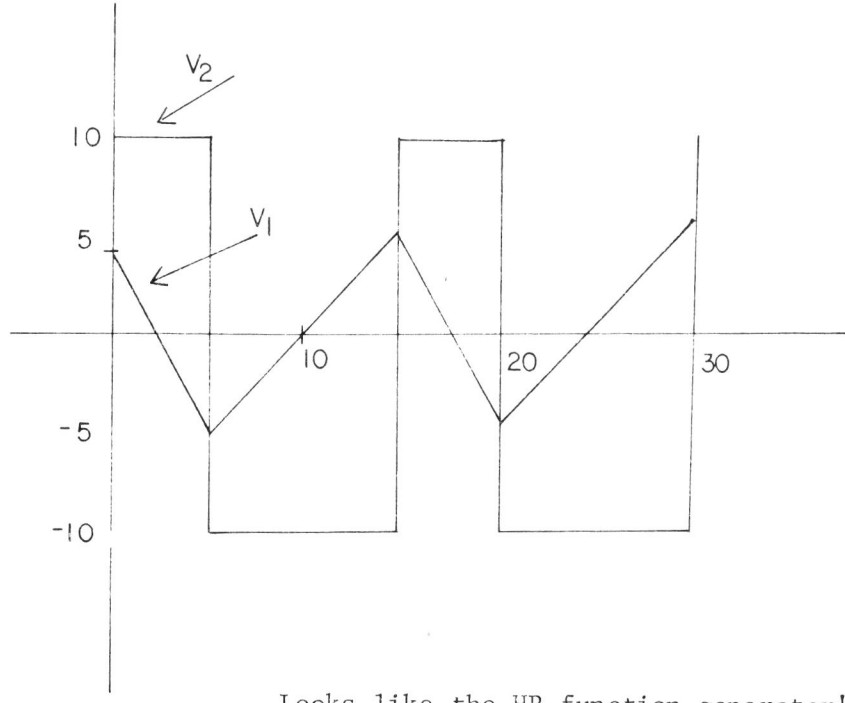

Looks like the HP function generator!

220 / THIRD SAMPLE P & P/EE EXAMINATION

Solution 3:

a) First convert the short-circuit data to per unit on 1,000 kva per phase.

For primary:
$$V_{base} = 7,960 \text{ volts}$$
$$I_{base} = \frac{1,000}{7.96} = 125.4 \text{ amp}$$

For secondary:
$$V_{base} = 2,400 \text{ volts}$$
$$I_{base} = \frac{1,000}{2.4} = 416 \text{ amp}$$

Conversion of the test data to per unit then gives

Test	Windings	V	I
1	1 and 2	0.0316	0.500
2	1 and 3	0.0967	0.500
3	2 and 3	0.0905	0.500

From test 1, the short-circuit impedance Z_{12} is

$$Z_{12} = \frac{0.0316}{0.15} = 0.0632 \text{ per unit}$$

Similarly, from tests 2 and 3,

$$Z_{13} = \frac{0.0967}{0.500} = 0.1934 \text{ per unit}$$

$$Z_{23} = \frac{0.0905}{0.500} = 0.1910 \text{ per unit}$$

The equivalent constants are

$$Z_1 = jX_1 = j0.0378 \text{ per unit}$$

$$Z_2 = jX_2 = j0.0254 \text{ per unit}$$

$$Z_3 = jX_3 = j0.1556 \text{ per unit}$$

b) Base line to line voltage for the Y-connected primaries is

$\sqrt{3} \, (7,960) = 13,800$ volts, or the bus voltage is 1.00 per unit. From the equivalent circuit with a short circuit on the tertiaries,

$$I_{sc} = \frac{V_1}{Z_1+Z_3} = \frac{V_1}{Z_{13}} = \frac{1.00}{0.1934} = 5.18 \text{ per unit.}$$

Solution 3 cont'd

(Note, however, that this current is 10.35 per unit on the rating of the tertiaries.) If the voltage drops caused by the secondary load current are neglected in comparison with those due to the short-circuit current, the secondary terminal voltage equals the voltage at the junction of the three impedances Z_1, Z_2, and Z_3, whence

$$V_2 = I_{sc} Z_3 = (5.18)(0.1556) = 0.805 \text{ per unit}$$

Solution 4:

a) $A_{v_2} \simeq \dfrac{R_L}{r_{in_2}} = \dfrac{2K}{40} = 50$

$A_{v_1} = \dfrac{-h_{fe} R_L}{h_{ie}} = \dfrac{-100 \times 40}{1.5K} = -2.67$

$A_v = A_{v_1} A_v = -2.67 \times 50 = -133$

b) $C_{in_1} = C_{b'e} + (1+|A_{v_1}|)C_{b'c} = 8\text{pf} + (1+|-2.67|)2\text{pf} = 15.3\text{pf}$

$C_S = C_{in_1}$ since C_W is negligible

$R_{TH} = R_g' || R_B || r_{in_1} = 1.5K$

$f = \dfrac{1}{2\pi C_S R_{TH}} = \dfrac{.159}{15.3\text{pf} \times 1.5K} = 6.9\text{MHz}$

c) $A_i \text{BW} = h_{fe} f = 100 \times 6.9\text{MHz} = 0.69\text{MHz}$

Solution 5:

$G = \dfrac{K(s+5)}{s(s+1)(s+10)} = \dfrac{K(s+5)}{s^3 + 11 s^2 + 10 s}$

Solution 5 cont'd

a) $\dfrac{C(s)}{R(s)} = \dfrac{G}{1+G} = \dfrac{K(s+5)}{s^3+11s^2+10s+K(s+5)}$

$\dfrac{C(s)}{R(s)} = \dfrac{K(s+5)}{s^3+11s^2+(10+K)s+5K}$

b) $\dfrac{C(s)}{R(s)} = \dfrac{Ks+5K}{s^3+11s^2+(10+K)s+5K}$

$= \dfrac{K/s^2 + \dfrac{5K}{s^3}}{1+11/s+\dfrac{(10+K)}{s^2}+\dfrac{5K}{s^3}}$

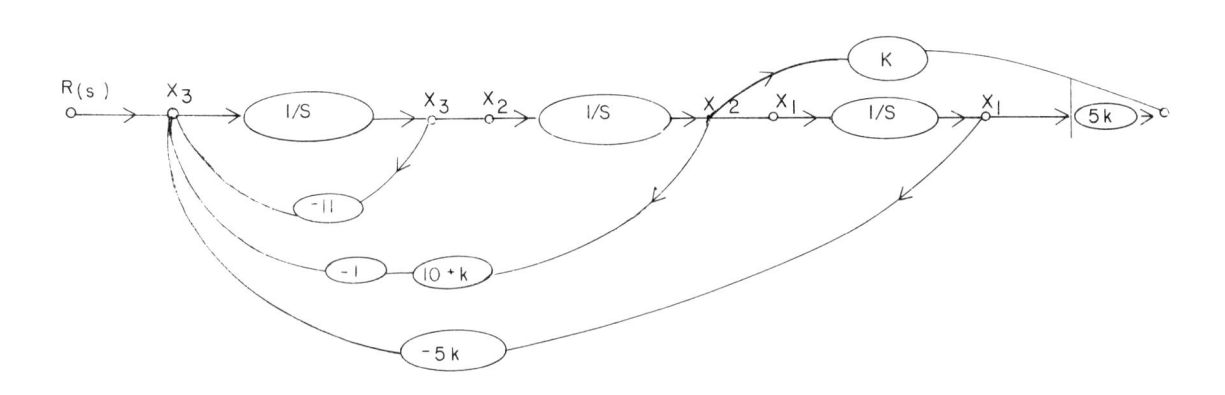

c) $\dot{X}_3 = -11X_3 - (10+K)X_2 - 5X_1 + r$ $\qquad$ $C = KX_2 + 5KX_1$

$\dot{X}_2 = X_3$

$\dot{X}_1 = X_2$

$\dot{\overline{X}} = \begin{bmatrix} 0 & 1 & 0 \\ 0 & 0 & 1 \\ -5K & -(10+K) & -11 \end{bmatrix} \begin{bmatrix} X_1 \\ X_2 \\ X_3 \end{bmatrix} + \begin{bmatrix} 0 \\ 0 \\ 1 \end{bmatrix} [r]$

Solution 5 cont'd

$$\overline{X} + \overline{A}\,\overline{X} + \overline{B}\,r$$
$$C = \overline{C}\,\overline{X} + \overline{D}\,r$$

$$C = \begin{bmatrix} 5K & K & 0 \end{bmatrix} \begin{bmatrix} X_1 \\ X_2 \\ X_3 \end{bmatrix} + \begin{bmatrix} 0 \end{bmatrix} \begin{bmatrix} r \end{bmatrix}$$

Solution 6:

a) Maximum demand for each apartment:

General-lighting load	2400 watts
Small appliance load	3000 watts
Total less range	5400 watts

Application of demand factors:

3000 watts at 100%	3000 watts
2400 watts at 35%	840 watts
Range load	8000 watts
Total demand load	11,840 watts

Current for each line conductor = $\dfrac{11840}{230}$ = 51.5 amps

Neutral load:

Lighting and appliance load - 3000 + 840	3840 watts
Range load - 8000 watts at 70%	5600 watts
Total neutral demand	9440 watts

Current for neutral conductor = $\dfrac{9440}{230}$ = 41 amps

Recommended answers:

    2 - #4 TW copper conductor
    1 - #6 TW copper neutral
or 2 - #6 RHW or THW copper conductors
    1 - #8 RHW or THW copper neutral

(Answers based on 40°C ambient - IPCEA ampacities. #6 TW - 44a, #4 TW - 59a, #6 RHW - 57a)

Acceptable answers but suggest reduced credit:

    3 - #6 TW copper conductors (NEC - 30°C ambient ratings)

224 / THIRD SAMPLE P & P/EE EXAMINATION

Solution 6 cont'd

b) Maximum demand for feeder supplying building:

Total loads:
    Lighting and small-appliance loads, 20 x 5400 = 108,000 watts

Application of demand factors:

3000 watts at 100%	3,000 watts
105000 watts at 35%	36,750 watts
Subtotal	39,750 watts
Range load, 20 ranges (less than 12kw)	35,000 watts
Total demand load	74,750 watts

Demand current for each line conductor = $\dfrac{74,750}{230}$ = 325 a

Neutral load:

Lighting and small appliance loads	39,750 watts
Range load - 35,000 at 70%	24,500 watts
	64,250 watts

Current demand for neutral = $\dfrac{64,250}{230}$ = 279 a

Further demand factor:

200 amps at 100%	200 a
79 amps at 70%	55.5 a
Max. current demand-neutral	255.5 a

Recommended answers: 2 - #500 MCM-RH or THW copper conductors
                                  1 - #350 MCM-RH or THW copper conductor
                 or   2 - #350 MCM-RHH or THHN copper conductors
                                    1 - #250 MCM-RHH or THHN copper conductor

(Answers based on 40 C ambient - IPCEA Ratings, 500 MCM TH-336a, 350 MCM RH-274, 350 MCM RHH-336 a)

Answers acceptable but suggest reduced credit:

    2-#500 MCM TV copper
    1-#350 MCM TW copper   (See note below)

(NEC - 30 C ambient ratings)

NOTE: In large wire sizes, (above 1/0) TW or R conductors are not not stocked. Common practice is to use RHW, THW or RHH, THHN types. Also, many sizes are not used, i.e., 300 or 400 MCM.

Solution 7:

a) $V_{DS} = \dfrac{V_{D(max)}}{2} = \dfrac{16}{2} = 8V$

$R'_L = \dfrac{V_{DS(max)} - V_{DS}}{I_D} = \dfrac{16-8}{20\text{ma}} = 400$

b) $n = \left(\dfrac{R'_L}{R_L}\right)^{1/2} = \left(\dfrac{400}{10}\right)^{1/2} = 6.3:1$ Primary to secondary turns ratio

c) $P_o = \dfrac{(11-3.8)V \times (31-12)\text{ma}}{8} = 17.2\text{mW}$

d) $P\text{ insert} = \dfrac{V_{gs}^2}{8R'_L} = \dfrac{(1V)^2}{8 \times 400} = 0.312\text{mW}$

Insertion $A_P = \dfrac{P_o}{P_{ins}} = \dfrac{17.2\text{mW}}{0.312\text{mW}} = 55$

e) % distortion $= \left| \dfrac{(i_{dmax} + i_{dmin})/2 - I_D}{i_{dmax} - i_{dmin}} \right| \times 100$

P.D. 2nd HAR $= \left| \dfrac{(31+12)\text{mA}/2 - 20\text{ma}}{(31-12)\text{ma}} \right| \times 100 = 7.9\%$

f) $r_d = \dfrac{\Delta V_{DS}}{\Delta I_D} = \dfrac{16V}{4\text{ma}} = 4K$; $g_f = \dfrac{\Delta I_D}{\Delta V_{GS}} = \dfrac{22\text{ma}}{1} = 22 \times 10^{-3}$

@ $V_{DS} = 8V$, $I_D = 20\text{ma}$

$P_o = \dfrac{(g_f v_{gs} r_d)^2 R'_L}{8(r_d + R'_L)^2} = \dfrac{(22 \times 10^{-3} \times 1 \times 4K)^2 \, 400}{8(4K+400)^2} = 20\text{mW}$

Solution 8:

a) $4\dfrac{di}{dt} + 2i = 4t \qquad \dfrac{di}{dt} = \tfrac{1}{2}i = t$

I. Transient Sol.

$\dfrac{di}{dt} = \tfrac{1}{2}i = 0 \qquad L\dfrac{di}{dt} + iR = 0 \qquad \dfrac{di}{dt} + \dfrac{R}{L}i = 0$

$$\tau = \dfrac{L}{R}$$

$\tau = 2$

$i_C = A\,-t/2$

II. Particular Sol.

$i_P = Bt+C \qquad \dfrac{di}{dt} = B$

Substitute $\quad B + \tfrac{1}{2}(Bt + C) = t$

$\qquad\qquad B + \tfrac{1}{2}Bt + \tfrac{1}{2}C = t$

$\qquad\qquad B + \tfrac{1}{2}C = 0 \qquad\qquad \tfrac{1}{2}Bt = t$

$\qquad\qquad\qquad\qquad\qquad\qquad\qquad B = 2$

$\qquad\qquad 2 - \tfrac{1}{2}C = 0 \qquad\qquad C = -4$

$i_P = 2t-4$

III. $i(t) = i_C + i_P$

$i(t) = A\varepsilon^{-t/2} + 2t - 4$

$i(0) = -3 = A + 0 - 4$

$\qquad\qquad A = 1$

$i(t) = \varepsilon^{-t/2} + 2t - 4$

b) $0 = \varepsilon^{-t/2} + 2t - 4$

$\qquad t = 1.797$ secs by iterative process

Solution 9:

a) $P_{R_L} = E_L I \cos\phi = 100I(.8) = 80I$

$Q_L = E_L I \sin\phi = 100I(.6) = 60I$

$P_1 = I^2 \cdot 1 = I^2 \qquad\qquad Q_1 = I^2 \cdot 1 = I^2$

$P = (P_1 + P_{R_L}) + j(Q_1 + Q_L) = I^2 + 80I + j(I^2 + 60I)$

$P = VI = 120I = I^2 + 80I + j(I^2 + 60I)$

$\qquad 120 = I + 80 = j(I + 60)$

$(120)^2 = (I+80)^2 + (I+60)^2$

$\qquad I = 14.26 \text{ amps}$

$\theta = \arctan \dfrac{I+60}{I+80} = \arctan \dfrac{74.26}{94.22} = 38.23° \qquad\qquad \theta = 38.23°$

$\qquad\qquad\qquad\qquad\qquad\qquad\qquad\qquad\qquad\qquad P_f = \cos\theta = 0.786$

b) $P_T = 100I(.5) = IE_L(.8)$

$\qquad E_L = 62.5V$

$Q_T = 100I(.866) = I^2(1) + I(62.5)(.6)$

$\qquad I = 49.1$

$X_C = \dfrac{E}{|I_m I|} \qquad ; \qquad \theta = \arccos \tfrac{1}{2} = 60$

$\qquad\qquad\qquad\qquad I_Q = 49.1 \sin 60 = 42.5$

$X_C = \dfrac{100}{42.5} = 2.353\,\Omega$

AFTERNOON SECTION

You will have four hours in which to work this test. Your score will be directly proportional to the number or problems you solve correctly through four (4). Each correct solution counts ten points. The maximum possible score for this part of the examination is 40 points. Partial credit for partially correct solutions will be given.

Work four of the problems according to instructions. Do not submit solutions or partial solutions for more than four problems. Indicate the problems which you have solved.

You may work only one engineering economy problem. When you have completed this portion of the P&P examination in the required time limit, you should check your solutions with the answers at the end of this examination. For additional practice you are encouraged to work the other problems under the time limit.

Since you want the maximum points available, you should remember that the examiner who assigns these points must make his judgment based only on what has been written down during the examination. It is important for you to be reasonably neat in your work and write down any assumption that you consider necessary to allow you to solve the problem properly and to provide sufficient rationale so that the examiner can judge your reasoning. Assumptions should follow the logic and requirements of the problem.

You are advised to use your time effectively.

1. An investor purchased 100 shares of stock in the Ajax Mutual Fund for $2,000. He received no dividends the first year, $1.75 a share dividends at the end of each year for the next 4 years, and $2.50 a share at the end of each of the 5 following years. After holding the stock for 10 years, he sold it for $5,500. What was the true rate of return on his investment?

2. The following waveshape is generated in a silicon controlled rectifier circuit.

    a) Find the average value.
    b) Find the RMS value.

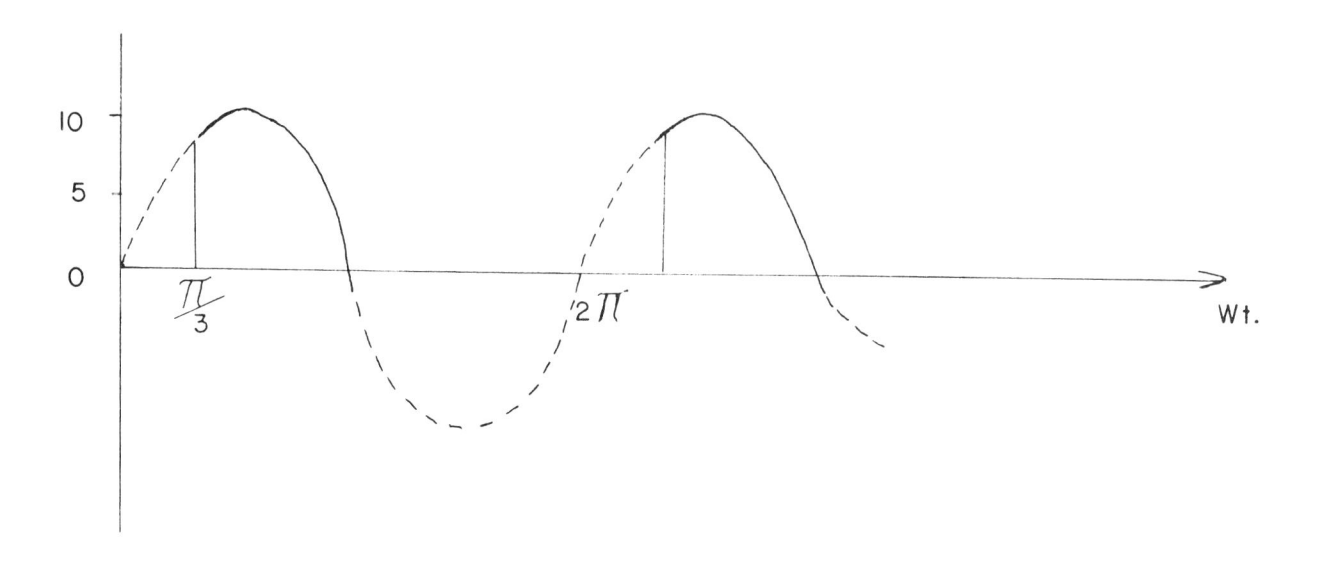

3. A three-phase, 13,000-v, 2500-kva alternator has an armature resistance of 0.3Ω/phase and a synchronous reactance of 4.0 ohms/phase. The alternator excitation is adjusted in each of the cases to provide rated voltage at rated load. If, after the adjustment is made, the load is suddenly removed from the alternator, calculate the no-load voltage per phase and line for loads of

   a) Unite power factor
   b) 0.8 lagging
   c) 0.8 leading
   d) Calculate voltage regulation for each of the above and determine best regulation.

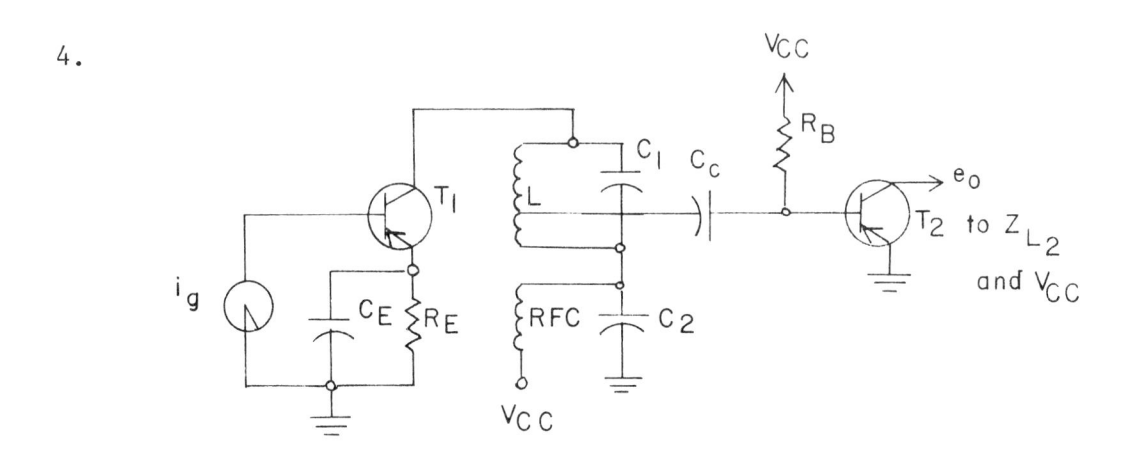

4.

A tapped, single, tuned circuit is to be used as the coupling method between two CE transistor amplifiers, as shown. The following circuit values and conditions are known:

$$r_{o_1} = 10K\Omega, \quad r_{in_2} = 2K\Omega$$

The unloaded inductor $L(=L_p)$ is to have a Q of 500, and must be resonant at 5MHz with $C(=C_p)$ to provide a bandwidth of 50kHz. Find the following values:

   a) Loaded tuned-circuit Q_o
   b) Coil inductance
   c) Number of turns required for the inductor single-layer wound on an air core of ¼ in. radius and 1 in. in length.
   d) Approximate tap locations
   e) Capacitance required for the desired resonance.

5. In the figure shown, G is defined by $0.004 \frac{dc}{dt} + c = 80e$ and H is defined by $0.002 \frac{df}{dt} + f = Kc$. Write the differential equation relating c and r.

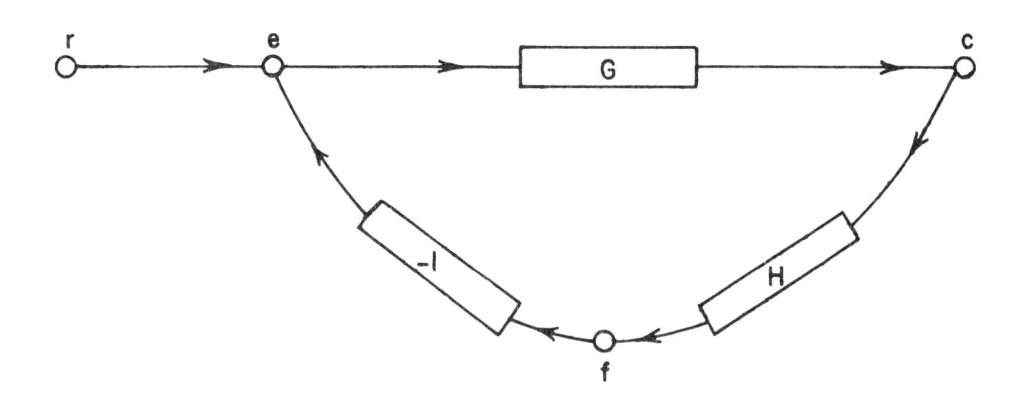

6. Make an economic study with respect to a lighting installation to provide a maintained illumination of 30 ft. c. in a 15,000 sq. ft. warehouse. Assume an electrical energy cost of 75 mils flat rate and operation of 3,120 hours per year. A CU of 80% and M.F. of 70% is assumed to apply to both types of luminaires. The luminaire (fixture) costs given includes lamp cost and material and labor for installation of transformers, ballast, panels, conductors, conduit, and auxiliary equipment as required - all pro-rated in the given figure. Lamp replacement cost includes lamp, installation, and cleaning cost. The systems to be compared use luminaires as follows:

   Fixture A: Incandescent - 150 watt, 115 volt, initial lumens 2600, life 750 hours, fixture cost $60, lamp replacement cost $3.50

   Fixture B: Mercury vapor - 175 watt, 115 volt, initial lumens 7700, life 24,000 hours, fixture cost $135, lamp replacement cost $8.00, ballast loss 25 watts.

Compare both installation cost and annual cost of operation.

7. A drain characteristic for a P-channel junction FET is shown below. Determine the following for a common source amplifier.

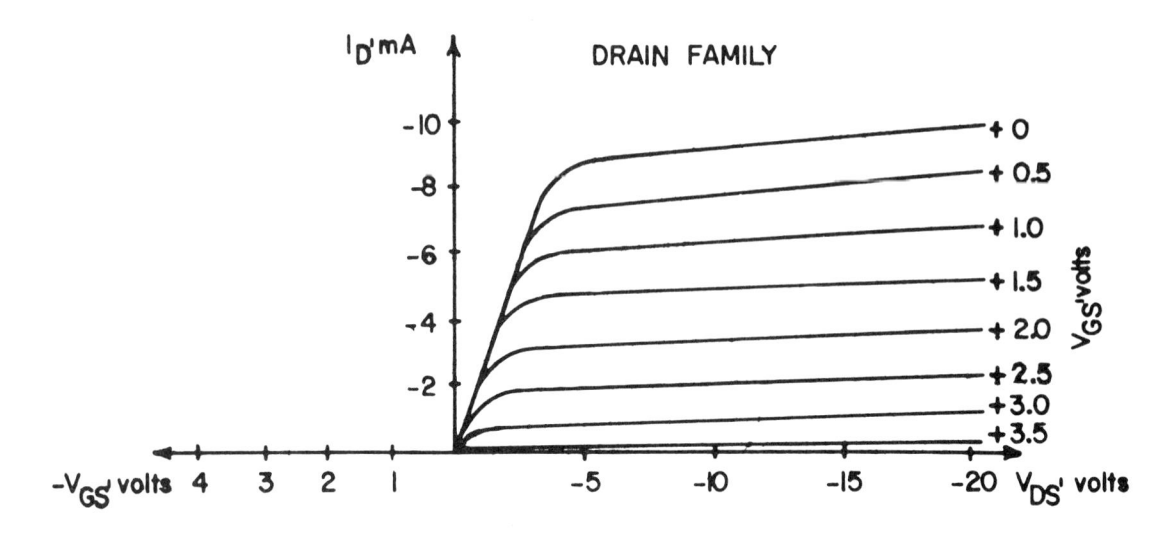

a) The plot of $R_{L_{ac}}$ = 2KΩ when the Q-point is identified as
   V_{DS} = -11.5V, I_D = -3.8 mA, and V_{GS} = +2V

b) R_S source resistor, to obtain Q-point by self-biasing.

c) C_S for adequate R_S bypass down to 100 H_z.

d) Determine the necessary V_{DD} supply value.

e) Given a v_{gs} = 2V_{pp} find the voltage gain of this amplifier.

f) Find r_D and g_f at the Q-point.

g) Find A_v from device parameters and compare with e.

8.

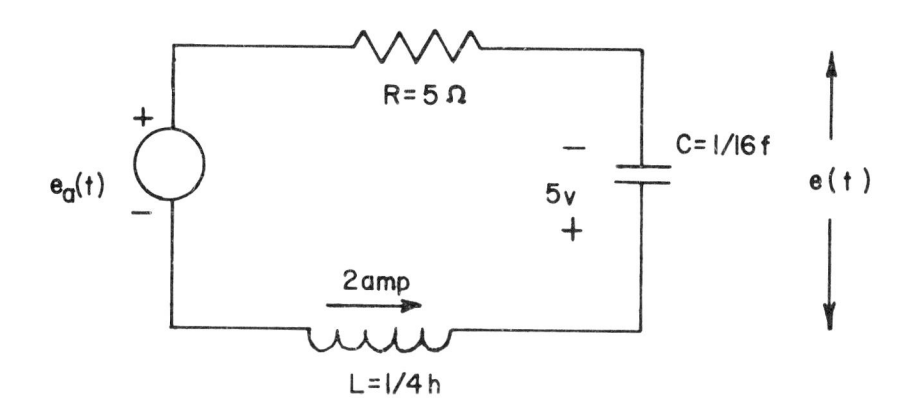

[Circuit diagram: $e_a(t)$ source on left with + top, − bottom. Top branch has R = 5 Ω resistor. Right branch has capacitor C = 1/16 f with − on top, + on bottom, marked 5v. Bottom branch has inductor L = 1/4 h with 2 amp current flowing right. Output e(t) measured across capacitor on the right.]

Given that $e_a(t) = 12v$ for $t \leq 0$ find $e(t)$ across the capacitor.

9. The circuit shown represents the equivalent circuit for a transformer. It includes the winding resistances and an extra resistance R_m to account for the energy lost in the iron core. The values of X_m and R_m are much larger than the other elements in the circuit.

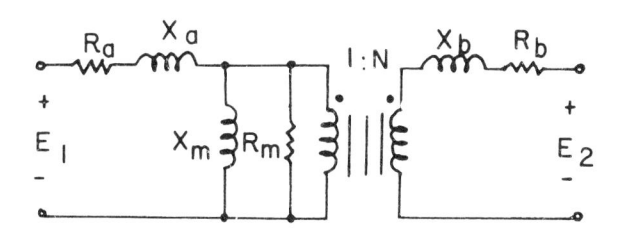

[Circuit diagram: Transformer equivalent circuit with input E_1 on left through R_a and X_a in series, then parallel branch with X_m and R_m, through ideal transformer 1:N, then X_b and R_b in series to output E_2.]

(a) Find the input impedance when the output is short-circuited. Obtain an approximate value which utilizes the conditions stated above.

(b) Find the input impedance with the output open-circuited. Again use reasonable approximations.

(c) A 10-KVA, 60-cps, 2300/230-volt distribution transformer is tested with the secondary open and with the secondary shorted. The data are as follows:

    Open-circuit data
      Input voltage: 2300 v (rated voltage)
      Input current: 1/10 amp
      Phase angle: 45° lagging current

9 cont'd

    Short-circuit data
      Input voltage: 100 v
      Input current: 100 amp (rated current)
      Phase angle: 45° lagging current

Assume that the leakage reactances and the winding resistances are equal when referred to the same side of the transformer. Obtain an equivalent circuit for the transformer referred to the high-voltage winding.

Solution 1:

$$2{,}000(\text{crf}) = 175\,(\text{uspwf})(\text{sppwf})(\text{crf}) + 250\,(\text{uspwf})(\text{sppwf})(\text{crf})$$
$$\quad\quad\text{i-10} \quad\quad\quad\quad \text{i-4}\quad\quad \text{i-1}\quad\quad \text{i-10} \quad\quad\quad\quad\quad \text{i-5}\quad\quad \text{i-5}\quad\quad \text{i-10}$$

$$+\; 5{,}500(\text{sfdf}),\; \text{whence } i = 16.5\%$$
$$\quad\quad \text{i-10}$$

Solution 2:

$$\omega = \frac{2\pi}{T} = \frac{2\pi}{2\pi} = 1$$

a) $f(\text{ave}) = \dfrac{1}{T} \displaystyle\int_0^{2\pi} f(t)\,dt = \dfrac{1}{2\pi} \displaystyle\int_{\pi/3}^{\pi} 10 \sin t \, dt$

$$= \frac{5}{\pi}(-\cos t)\Big|_{\pi/3}^{\pi} = \frac{5}{\pi}\left(-\cos\pi + \cos\frac{\pi}{3}\right)$$

$$= \frac{5}{\pi}(-(-1)+\tfrac{1}{2}) = \frac{5}{\pi}(1+\tfrac{1}{2}) = \frac{5}{\pi}\cdot 3/2 = \frac{15}{2\pi}$$

$$f(\text{ave}) = \frac{15}{2\pi} = 2.39$$

b) $f_{\text{rms}} = \sqrt{\dfrac{1}{T}\displaystyle\int_0^T f(t)^2 dt} = \sqrt{\dfrac{1}{2\pi}\displaystyle\int_{\pi/3}^{\pi}(10\sin t)^2 dt}$

$$= \sqrt{\frac{100}{2\pi}\int_{\pi/3}^{\pi}\sin^2 t\,dt} = \sqrt{\frac{50}{\pi}\int_{\pi/3}^{\pi}\left(\tfrac{1}{2}-\frac{\cos 2t}{2}\right)dt} = \sqrt{\frac{50}{\pi}(\tfrac{1}{2}t)\Big|_{\pi/3}^{\pi} - \frac{\sin^2 t}{4}\Big|_{\pi/3}^{\pi}}$$

$$= \sqrt{\frac{50}{\pi}(\tfrac{1}{2}(\pi - \pi/3) - \tfrac{1}{4}(\sin 2\pi - \sin 2\pi/3)} = \sqrt{\frac{50}{\pi}(\pi/3 - \tfrac{1}{4}(0 - .866))}$$

$$= \sqrt{\frac{50}{\pi}(\pi/3 + \tfrac{.866}{4})} = \sqrt{\frac{50}{\pi}(1.26)} = \sqrt{20.05} = 4.48$$

$$f_{\text{rms}} = 4.48$$

Solution 3:

$$I_L = I_P = \frac{kva}{\sqrt{3}V_L} = \frac{2500.0}{1.73 \times 13.0} = 111a. \qquad V_P = \frac{V_L}{\sqrt{3}} = \frac{13kv}{1.73} = 7.51kv$$

$V_L = 13kv$ $\qquad\qquad\qquad\qquad$ $I_P R_a = 111 \times 0.3 = 33.3v$

$2500 kva$ $\qquad\qquad\qquad\qquad$ $I_P X_S = 111 \times 4 = 444v$

$R_a = 0.3\Omega/\phi$

$X_s = 4\Omega/\phi$ $\qquad\qquad$ a) at <u>unity</u> p.f.

$$E_{gp} = V_P + I_a R_a + jI_a X_s = 7510 + 33.3 + j444 = 7543 + j444$$
$$= 7543.3v$$

b) At 0.8 p.f. lagging

$$E_{gp} = \left[7510 \times 0.8 + 33.3\right] + j\left[7510 \times 0.6 + 444\right] = 6041 + j4950 \quad \underline{/39.4}$$
$$= 7800v$$

c) At 0.8 p.f. leading

$$E_{gp} = \left[7510 \times 0.8 + 33.3\right] + j\left[7510 \times 0.6 - 444\right] = 6041 + j4602 = \underline{/33.9}$$
$$= 7275v$$

d) VR, at unity p.f. $= \frac{7543 - 7510}{7510} \times 100 = 0.439\%$

VR, at 0.8 p.f. lagging $= \frac{7800 - 7510}{7510} \times 100 = 3.86\%$

VR, at 0.8 p.f. leading $= \frac{7275 - 7510}{7510} = -3.13\%$

best regulation occurs at unity p.f. - with zero percent at slightly leading (i.e., between unity and approx 0.9 p.f.)

Solution 4:

a) $Q_O = \frac{f_O}{BW} = \frac{5MHz}{50KHz} = 100$

b) $L_P = \frac{r_O}{W_O} \left(\frac{1}{2Q_O} - \frac{1}{Q}\right) = \frac{10K}{2\pi \times 5MHz}\left(\frac{1}{2 \times 100} - \frac{1}{500}\right) = .96\mu h$

Solution 4 cont'd

c) $n = \left[\dfrac{L_p(9r_r+10\ell)}{r}\right]^{1/2} = \left[\dfrac{.96(9\times.25+10\times1)}{.25}\right]^{1/2} = \dfrac{(11.7)^{1/2}}{0.25} = 13.7$

d) $N_2 = \left(\dfrac{r_{in_2}}{r_o'}\right)^{1/2} \times n = \left(\dfrac{2K}{6.1K}\right)^{1/2} \times 43 = 8$ turns from ground side of coil

where $r_o' = 2W_o L_p Q_o = 6100$

e) $C = \dfrac{1}{4\pi^2 f^2 L} = \dfrac{1}{4\pi^2(5\text{Mhz})^2 \times .96\mu h} = .001\mu f$

Solution 5:

$G \rightarrow 0.004 \dfrac{dc}{dt} + c = 80e$

$H \rightarrow 0.002 \dfrac{df}{dt} + f = Kc$

$.004sc(s) + c(s) = 80\, E(s)$

$c(s)\,(.004s+1) = 80\, E(s)$

$G(s) = \dfrac{c(s)}{E(s)} = \dfrac{80}{.004s+1} = \dfrac{80}{.004(s + \frac{1}{.004})}$

$G(s) = \dfrac{20000}{s+250}$

$H(s) = \dfrac{F(s)}{c(s)} \qquad .002sF(s) + F(s) = Kc(s)$

$\dfrac{F(s)}{c(s)} = \dfrac{K}{.002s+1}$

$H(s) = \dfrac{K}{.002} \cdot \dfrac{1}{s+\frac{1}{.002}}$

$H(s) = \dfrac{500K}{s+500}$

$\dfrac{c(s)}{r(s)} = \dfrac{G}{1+GH}$

$\dfrac{c(s)}{r(s)} = \dfrac{\dfrac{20000}{s+250}}{1 + \left(\dfrac{20000}{s+250}\right)\left(\dfrac{500K}{s+500}\right)}$

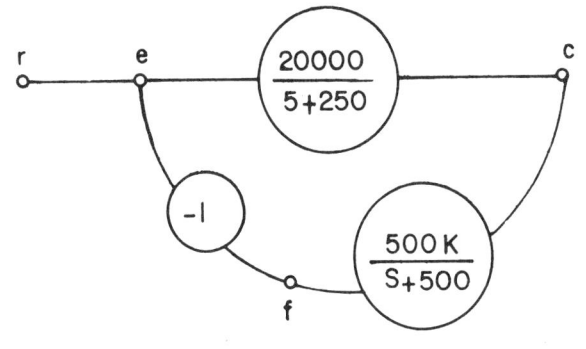

Solution 5 cont'd

$$\frac{c(s)}{r(s)} = \frac{2000(s+500)}{(s+250)(s+500)+20000(500K)} = \frac{20000(s+500)}{s^2+750s+(250)(500)+20000(500K)}$$

$$s^2 c(s) + 750x\, c(s) + K_1 c(s) = 20000 s R(s) + 10^6 R(s) \qquad K_1 = 750000 + 10^6 K$$

$$\frac{d^2 c}{dt^2} + 750\frac{dc}{dt} + K_1 c = 20000\frac{dr}{dt} + 10^6 r$$

Solution 6:

No. of luminaires = $N = \dfrac{EA}{\phi(CU)(MF)}$

Incandescent (Fixture A):
$N = \dfrac{(30)(15,000)}{(2600)(0.8)(0.7)} = 309$

Initial cost = (309)($60) = $18,540

Annual operating cost:

    Energy = (0.150)(309)(3120)(0.075) = $10,486

    Lamp   = (3.50)(309)(3120/750)   = $\underline{4,500}$

                                                                 $15,346

Mercury Vapor (Fixture B):
$N = \dfrac{(30)(15,000)}{(7700)(0.8)(0.7)} = 105$

Initial cost = (105)(135) = $14,175

Annual operating cost:

    Energy = (0.200)(105)(3120)(0.075) = $4,914

    Lamp   = (105)(8)(3120/24000)   = $\underline{109}$

                                                                  $5,023

Solution 7:

a) X axis intercept of $R_{L_{ac}} = V_{DS} + I_D R_{LC}$ = $-11.5 - 3.8\text{ma} \times 2K$
                                                                                                        = -19.1V

Solution 7 cont'd

b) $R_S = \dfrac{V_{GS}}{I_D} = \dfrac{2.0}{3.8mA} = 526$

c) $X_{CS} = 0.1 R_S = 0.1 \times 526 = 52.6$

$f_{min} = 100 h_z$

$C_S = \dfrac{1}{2\pi f_{min} \times X_{CS}} = \dfrac{1}{2\pi \times 100 \times 52.6} = 30\mu f$

d) $R_{L_{DC}} = R_{L_{AC}} + R_S = 2K + 475 = 2475$

$V_{DD} = V_{DS} + I_D R_{L_{DC}} = -11.5 + (-3.8ma) \times 2475 = -20.9V$

e) $A_v = \dfrac{\Delta v_{ds}}{\Delta v_{gs}} = \dfrac{-17-(-5)}{2.8-0.8} = -6$

f) $r_d = \dfrac{\Delta V_{DS}}{\Delta I_D}\bigg|_{V_{GS}} = \dfrac{10}{0.3 mA} = 33K$

$g_f = \dfrac{\Delta I_D}{\Delta V_{GS}}\bigg|_{V_{DS}} = \dfrac{2ma}{0.5V} = 4000\mu$

g) $A_v = -g_f \dfrac{r_d R_L}{r_d + R_L} = -4000\mu \times \dfrac{33K \times 2K}{33K+2K} = -7.5$

Solution 8:

a) $\alpha = \dfrac{R}{2L} = \dfrac{5}{2 \cdot \frac{1}{4}} = 10$ $P_1 = \alpha - \sqrt{\alpha^2 - \omega_0^2} = 10 - \sqrt{100-68} = 10-6 = 4$

$P_2 = \alpha + \sqrt{\alpha^2 \ \omega_0^2} = 10 + 6 = 16$

Solution 8 cont'd

b) $\omega_0 = \frac{1}{\sqrt{LC}} = \frac{1}{\sqrt{\frac{1}{4} \cdot 1/16}} = \frac{1}{\sqrt{1/64}} = 8$

c) Since $\alpha > \omega_0$ ω_d is irrelevant - circuit overdamped

d) $e(0) = -5$

e) $i_C = C\frac{de}{dt}$ $\left.\frac{de}{dt}\right|_0 = \frac{i(0)}{C} = \frac{-2}{1/16} = -32$

f) $e(\infty) = 12$

g&h) $i(t) = A_1 \varepsilon^{-k_1 t} + A_2 \varepsilon^{-k_2 t}$

$\frac{di^2}{dt} + 2\alpha\frac{di}{dt} + \omega_0^2 i = f(t)$

$\frac{di^2}{dt} + 20\frac{di}{dt} + 64 i = 12$

$k^2 + 20k + 64 = 0$

$k = \frac{-20 \pm \sqrt{400-256}}{2}$

$k = \frac{-20 \pm 12}{2}$

$k_1 = -4, \; k_2 = -16$

$i(t) = A_1 \varepsilon^{-4t} + A_2 \varepsilon^{-16t} \quad i(\infty) = 0$

$i(0) = -2 = A_1 + A_2$

$\left.\frac{1}{4}\frac{di}{dt}\right|_0 - 5 - 2(5) = 12$

$\left.\frac{di(t)}{dt}\right|_0 = 108 = -4A_1 - 16A_2$

$A_1 + A_2 = -2$

$-A_1 - 4A_2 = 27 \qquad -3A_2 = 25 \qquad A_2 = -25.3$

$A_1 - 25/3 = -2 \quad A_1 = +25/3 - 6/3 = 19/3$

$i(t) = 19/3 \varepsilon^{-4t} - 25/3 \varepsilon^{-16t}$

Solution 8 cont'd

i) $e(t) = 1/C \int i\, dt = 16 \int (19/3\varepsilon^{-4t} - 25/e\varepsilon^{-16t})\, dt$

$e(t) = \dfrac{16 \cdot 19}{3(-4)} \varepsilon^{-4t} + \dfrac{25 \cdot 16}{3 \cdot 16} \varepsilon^{-16t} + C$

$e(t) = -25.33\varepsilon^{-4t} + 8.33\varepsilon^{-16t} + C$

$e(\infty) = 12 \therefore C = 12$

$e(t) = -25.33\varepsilon^{-4t} + 8.33\varepsilon^{-16t} + 12$

Solution 9:

X_m and $R_m \gg X_a, X_b$ and R_a, R_b.

(a) Find $Z_{1_{SC}}$ with $E_2 = 0$

Transfer X_b and R_b

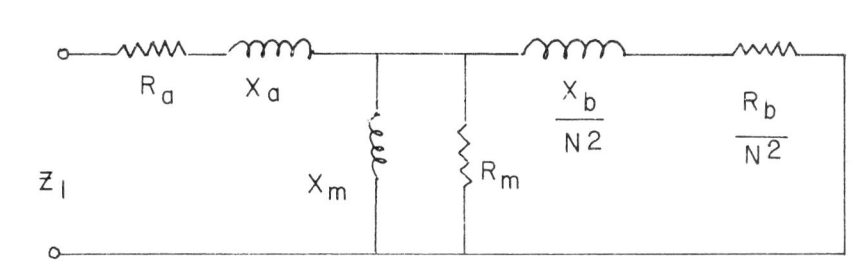

Since $X_m \gg X_b$
$R_m \gg R_b$ in parallel X_m and R_m can be ignored, assuming the turns ratio is not so great as to raise the values appreciably.

$\therefore Z_{1_{SC}} \simeq R_a + \dfrac{R_b}{N^2} + j(X_a + \dfrac{X_b}{N^2})$

b) Open Circuit: R_a and X_a are negligible.

$Z_{1_{OC}} = \dfrac{1}{\dfrac{1}{R_m} + \dfrac{1}{jX_m}}$

Solution 9 cont'd

c) 19KVA, 60Hz, 2300/23° O.C. $V_1 = 2300V$ S.C. $V_1 = 100V$

$I_1 = 1/10 a$ $I_1 = 100 a$

$\theta = 45°$ lag $\theta = 45°$ lag

1. O.C.

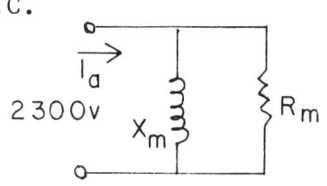

$Y = \dfrac{I}{V} = \dfrac{.1}{2300} \underline{|-45°} = \dfrac{.707}{23000} - j\dfrac{.707}{23000}$

$G = \dfrac{.707}{23000}$ $B_L = \dfrac{.707}{23000}$: $R = \dfrac{1}{G} = 32,600$

$X = \dfrac{1}{B_L} = 32,600$

$X_m = 32,600 \Omega$

$R_m = 32,600 \Omega$

2. S.C.

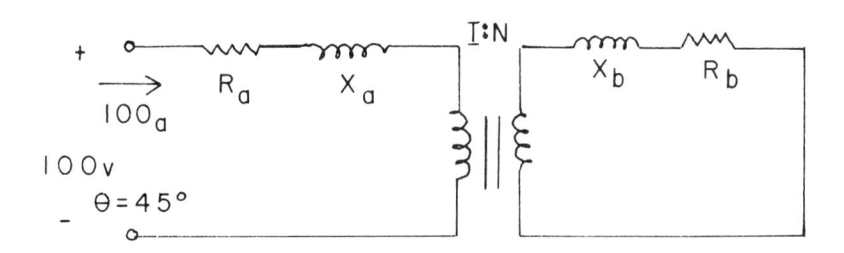

Assume:
$R_a = \dfrac{R_b}{N^2}$

$X_a = \dfrac{X_b}{N^2}$

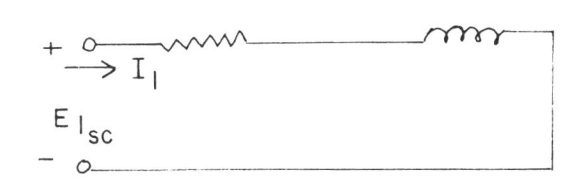

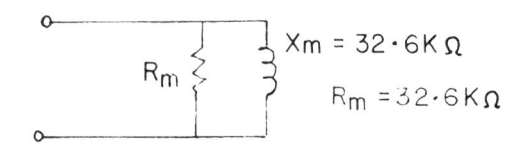

Solution 9 cont'd

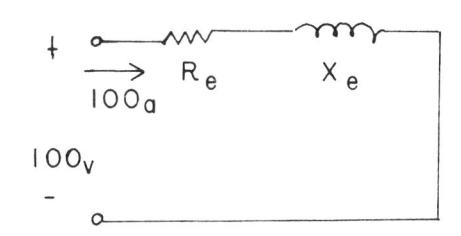

$$R_e = R_a + \frac{R_b}{N^2}$$

$$X_e = X_a + \frac{X_b}{N^2} \qquad N = \frac{1}{10}$$

$$R_e = R_a + 100 R_b$$

$$X_e = X_a + 100 X_b$$

by Voltage Division

$$E_{R_e} = 70.7V \qquad E_{X_e} = 70.7V$$

$$R_e = \frac{E_{R_e}}{I} = \frac{70.7}{100} = .707\Omega = X_e$$

or $Z_e = \frac{100}{100} = 1 \angle 45°$

$$Z_e = .707 + j.707$$

Now

$$R_a = 100 R_b = \frac{R_e}{2} = \frac{.707}{2} = .3535$$

$R_a = .3535\Omega$
$X_a = .3535\Omega$

$R_b = \frac{.3535}{100}$

$R_b = .003535\Omega$
$X_b = .003535\Omega$

$R_e \quad .3535\Omega \qquad X_e \quad .3535\Omega$

$X_m = 32.6 K\Omega$
$R_m = 32.6 K\Omega$

Solution 9 cont'd

2. S.C. cont'd

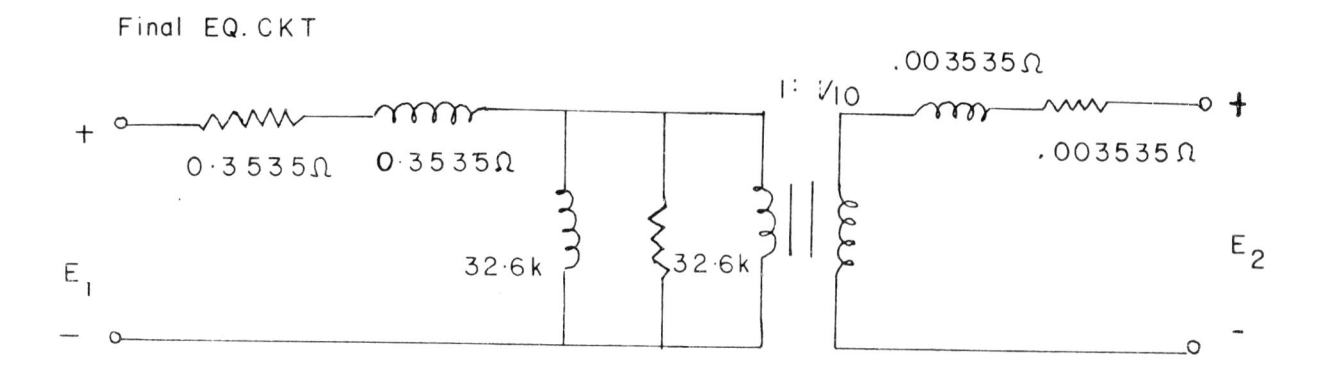

Final EQ. CKT

Part V
Appendices

Appendix 1

REVIEW:
FUNDAMENTALS OF ELECTRICAL GENERATORS AND MOTORS

Generators and alternators are widely used as a source of electrical power and motors are widely used as a source of mechanical power.

The characteristics of these machines have definite implications with regard to the design and construction of electrical equipment in general, so all of us in the field of electronics should know something about them.

This review presents a qualitative analysis of generators, alternators, dc motors, and ac motors.

At the end of each section you will find several review questions. These are study aids, and correct answers are provided.

1. <u>Electrical Generators</u>

The electrical generator is by far the most important source of electrical energy. It is a rotary-action machine which converts mechanical energy into electrical energy. Its size varies with the amount of electrical energy it is to produce. The smallest generator can produce a fraction of a watt of electrical power and can easily fit into the palm of one hand; the largest can produce a few hundred million watts and requires a multiple-story building to house it. A small generator may light a bicycle headlamp. A large one can supply enough energy for lighting an entire city.

The principle of generator action requires that magnetic lines of force be cut by conductors. The cutting of the magnetic lines may be done by keeping the magnetic field stationary and moving the conductors through the field, or by keeping the conductors stationary and moving the magnetic field about the conductors.

The operation of a generator is based on the fact that an emf will be induced in a conductor whenever the conductor cuts magnetic lines of force.

The main components of a generator are the field poles and the armature. The field poles are mounted on the housing of the generator and surround the armature so as to leave the minimum possible airgap between the pole faces and the surface of the armature.

The armature is the rotating portion of the generator. It consists of a shaft on which a laminated iron core and a commutator are mounted. The iron core has many radial slots at its outer surface. The many conductors which make up the armature winding fit into these slots. The commutator serves the dual purpose of rectifying the generated emf and conducting it through one or more pairs of brushes to the outer circuit.

The armature winding of the modern generator may be of the lap type or the wave type. With the lap winding, an element under one pole is connected to an element occupying a nearly corresponding position under the next pole. This second element is then connected back to an element under the original pole, but removed two or more elements from the initial one. With the wave winding, the second element, instead of being connected back to an element underneath the initial pole, is connected to a corresponding element underneath the next pole. The wave winding thus advances around the armature from pole to pole until all of the armature slots have been filled.

An advantage of the wave winding over the lap winding is that, for a given number of winding elements, the wave winding has a greater emf output than the lap winding. Hence, when a high voltage is required from a relatively small generator, we use a wave winding for the armature.

REVIEW QUESTIONS

1. State the fundamental principle of generator action.

2. What is the neutral zone of a generator?

3. In what plane is a conductor moving with respect to the stationary field at the instant that maximum emf is induced in the conductor?

4. What is the function of a commutator in a generator?

5. What is the advantage of a closed type of winding over an open type of winding?

6. What is the difference between a progressive and a retrogressive armature winding?

7. What determines the number of current paths in a lap-wound armature?

8. State an advantage of wave type winding over the lap type winding.

ANSWERS TO REVIEW QUESTIONS

1. The principle of generator action is that an emf is induced into a conductor whenever the conductor cuts magnetic lines of force.

2. The neutral zone is the zone in which the conductor is moving parallel to the magnetic lines of force and, consequently, generating no voltage.

3. At the instant that maximum voltage is being induced into a conductor it is moving in a plane perpendicular to the magnetic lines of force.

4. The purpose of the commutator in a generator is to rectify the emf induced into each coil of the armature winding.

5. The advantage of a closed type of winding over an open one is that in the former the individual coils of the winding are in series and are continuously active in generating emf.

6. In the progressive winding, each succeeding winding element is just ahead of the preceding element on the surface of the armature, while on the retrogressive winding each succeeding element is just behind the preceding element.

7. In a lap type winding there are as many current paths as there are field poles.

8. An advantage of a wave type winding over a lap type winding is that for the same number of winding elements the former yields a greater output voltage than the latter.

2. <u>Generator Characteristics</u>

Generators can be made to have a number of different characteristics, depending on the use for which they are intended. Thus we have shunt generators, series generators, and compound generators. Some generators may also be used as electromechanical amplifiers.

The emf induced in a conductor depends on the length of the conductor, the strength of the magnetic field in which it is moving, and the speed at which it is moving through the magnetic field. Since the conductors between any two brushes are in series with one another, the voltage existing between the same two brushes is the sum of the emf's induced in each of the conductors.

The saturation curve of a generator shows how the field strength varies with increasing field current. This curve determines in a large measure the characteristics of a shunt generator. In general, a high degree of magnetic saturation will prevent an excessive drop in terminal voltage as a result of an increase in load.

An important factor affecting the operation of a generator is armature reaction. Armature reaction results from the interaction of the main field with the field produced by current flowing through the armature. One of the effects of armature reaction is to shift the neutral plane in the direction of armature rotation by an amount that depends on the generator load. This requires that the brush axis be moved to the new or load neutral plane in order to prevent severe commutator sparking. Shifting the brush axis causes a corresponding shift in the axis of the field produced by the armature current. Since a component of this field is in opposition to the main field, one of the end results of armature reaction is to weaken the main field.

Sparking at the brushes occurs even when the brush axis is moved to the load neutral plane. The reason for this is that, when the individual coil windings of the armature reach the load neutral plane, the currents in the coils are still decaying and, in so doing, induce a back emf. We have called this back emf the "emf of self-induction". It is possible to neutralize this emf and, therefore, minimize sparking by advancing the brush axis to a point slightly ahead of the load neutral plane.

In the case of large generators or generators which do not operate under a constant load, the effects of armature reaction and of the emf of self-induction are both neutralized by the use of commutating poles. Since commutating poles are placed in the geometrical neutral planes and therefore neutralize the effect of armature reaction in these regions, there is no necessity for shifting the brush axis when the load is changed.

The ability of a generator to maintain a relatively constant voltage under varying loads is known as regulation.

The regulation of a shunt generator is poor. However, it can be considerably improved by using a field winding in series with the armature winding. When this is done, the generator becomes a compound generator. Compound generators can be wound so that the output voltage remains fairly constant with an increase in load. In this case, the generator is said to be flat compounded. The generator may also be wound so that the output voltage increases with an increase in load, in which case the generator is said to be over compounded. Flat-compounded generators are usually used when the load is near the generator, while over-compounded generators are usually used when the load is at a considerable distance from the generator.

The fact that a relatively small field current in a dc generator makes for a much larger current at the output suggests the use of the machine as an electromechanical amplifier. When so used, generators can supply power outputs far in excess of that available from vacuum-tube type amplifiers.

When one generator does not supply enough power, it is possible to obtain the desired amount of power by cascading two or more generators. When connected in cascade, the output of one generator provides the excitation current for the field winding of another generator.

An amplidyne is a specially designed generator capable of providing very large amounts of power amplification. The amplidyne has two sets of brushes. One set is in the usual position at right angles to the main field; these brushes are short-circuited. As a consequence, a large current flows through the armature, developing a strong field at right angles to the main field. As the armature conductors turn through this field, a voltage is induced. A second set of brushes placed at right angles to the first set makes this induced voltage available for driving an external load. Because the current available at the second set of brushes is proportional to the main field current, and as it may be several thousand times greater than the field current, the machine is capable of producing a very high order of power amplification. Amplidynes are widely used in radar for controlling the azimuth and elevation of large antennas.

REVIEW QUESTIONS

1. Name the factors which determine the magnitude of the emf induced in a conductor moving in a magnetic field.

2. What information does the saturation curve provide?

3. Name the three types of generators.

254 / ELECTRICAL GENERATORS AND MOTORS

4. Compare the field winding of a shunt generator with that of a series generator.

5. What is critical field resistance?

6. What is meant by "flashing the field"?

7. Mention two effects of armature reaction in a generator.

8. What is a cause of commutator sparking?

9. What is the function of commutating poles in a generator?

10. What term is used in describing the ability of a generator to maintain its voltage under varying load conditions?

11. How can a generator be made to have a greater output with increasing load?

12. What kind of generator would you recommend if the load is to be located at a considerable distance from the generator?

13. What characteristic of a generator permits its use as a power amplifier?

14. What determines the output power of an amplidyne?

15. What is the purpose of the compensating winding in an amplidyne?

ANSWERS TO REVIEW QUESTIONS

1. The strength of the magnetic field, the length of the conductor, and the speed at which it is moving.

2. It shows the manner in which the field strength of a generator increases with increasing field current.

3. Shunt, compound, and series.

4. The field winding of a series generator contains a much smaller number of turns and consists of a wire with a considerably larger cross section.

5. Critical field resistance is that value of field resistance above which a generator will fail to build up.

6. "Flashing the field" is the term used when an external voltage is applied to the field windings of a generator in order to start the voltage buildup of the generator.

7. One effect of armature is to weaken the field flux. Another effect is to shift the axis of the neutral plane.

8. Commutator sparking occurs if a voltage is induced in coils undergoing commutation.

9. The purpose of commutating poles is to neutralize the effect of armature reaction and emf of self-induction.

10. Generator regulation.

11. By placing a few turns of wire on the field poles which are in series with the armature.

12. An over-compounded generator.

13. The property of the generator whereby a small change in field current can produce a large change in output current suggests the use of the generator as a power amplifier.

14. For a given size amplidyne, the greater the input to the amplidyne the greater the output.

15. The purpose of the compensating winding in an amplidyne is to neutralize the effects of armature reaction.

3. The Motor

A generator is a machine that converts mechanical energy into electrical energy. On the other hand, a motor is a machine that converts electrical energy into mechanical energy. The two machines are very similar; in fact they are so similar that the "same machine" may be used either as a motor or as a generator.

When a generator is in operation, it is driven mechanically and develops a voltage, which, in turn, can produce a current flow in an electrical circuit; when a motor is in operation, it is supplied with current and develops torque, which, in turn, can produce mechanical rotation.

The electric motor is very similar in appearance and construction to any electrical generator. In fact, it could be said that the way the machine is used determines whether it is a motor or a generator.

In the case of a generator, the armature is mechanically driven and, as a result, develops a voltage. When a motor is in operation, a voltage from an external source is applied to the armature and, as a consequence, produces a torque.

The principle underlying the operation of an electric motor is that when a current is passed through a conductor lying in a magnetic field, the conductor experiences a force which causes it to move if it is free to move. The direction the conductor moves is such that direction of the magnetic field, direction of current through the conductor, and direction of motion of the conductor are all mutually perpendicular.

The armature of a dc motor, like that of a generator, has a winding consisting of a large number of coils. The large number of coils in this case makes for greater efficiency and smoothness of operation.

The rotation of the armature in a motor causes an emf to be induced into the armature in opposition to the applied voltage. This emf is known as counter emf and is the main factor limiting current in a motor. For this reason, it is necessary in the larger dc motors to apply voltage gradually to the armature. This is to give the armature time to pick up speed and develop sufficient counter emf before the full voltage is applied.

Armature reaction plays the same role in the motor as in the generator; it causes the neutral plane to shift. In motors, the brush axis must be retarded a few degrees from the no-load neutral plane to avoid sparking at the commutator.

In the motor as in the generator, the effects of armature reaction can be neutralized and the necessity for shifting the brush axis can be obviated through the use of commutating poles.

There are three types of dc motors, the series, the shunt, and the compound. The shunt and compound motors are most used because they have a definite no-load speed and lend themselves to speed control.

ELECTRICAL GENERATORS AND MOTORS

REVIEW QUESTIONS

1. Explain briefly the difference in the operation of a motor and that of a generator.

2. What does the torque of a motor depend on?

3. What is the main current-limiting factor in a motor?

4. How does the load neutral plane of a motor differ from that of a generator?

5. What is the distinguishing characteristic between the shunt motor and the series motor?

6. Why is it important that a series motor not operate without a load?

7. What advantage does a compound motor have over a shunt motor?

8. Why is it necessary to use a motor starter with the larger dc motors?

9. Explain two ways of controlling the speed of a motor.

10. How may the direction of rotation of a dc motor be reversed?

ANSWERS TO REVIEW QUESTIONS

1. With a generator, the armature is turned by a prime mover and a voltage is developed across the brushes. With a motor, an external voltage is applied to the brushes and as a result the armature turns.

2. The torque of a motor is proportional to armature current and the strength of the magnetic field.

3. The main current-limiting factor in a motor is counterelectromotive force.

4. In the generator, the load neutral plane is shifted counterclockwise, while in the motor, it is shifted clockwise (assuming that in both machines the armature turns in a clockwise direction).

5. A shunt motor is essentially a constant-speed motor. A series motor, on the other hand, has a higher starting torque but its speed drops off rapidly with increase in load.

6. A series motor should always be under load, otherwise it will run away.

7. A compound motor has a higher starting torque than a shunt motor.

8. With a large dc motor, the armature will not develop speed and, hence, counter emf in time to keep the armature from burning out due to excessive current.

9. The speed of a dc motor may be changed by changing either the applied voltage or the field strength.

10. The direction of rotation of a dc motor may be reversed by reversing the direction of current flow either through the armature or through the field winding.

4. The Alternator

The alternator is by far the most widely used source of electrical power in this country. Its name is derived from the fact that it delivers alternating current to the load instead of direct current, and its widespread use is due to the decided advantage of alternating over direct current, especially in the field of power distribution. Alternating current voltage may, by means of transformers, be stepped up to a value of many thousands of volts. At high voltages, the currents are very small, and for this reason, the power may be transmitted hundreds of miles with a minimum of I^2R loss. At the plant, where the electrical power is to be used, the voltage may be stepped down to any desired value. If a dc voltage is required, as in the case in many industrial plants, it can be readily obtained by rectifying the alternating-current voltage.

The alternator is a machine which converts mechanical energy into alternating-current electrical energy.

It differs from the dc generator in one important respect--the armature winding is held stationary and the field structure is rotated. With this arrangement it is possible to build a machine which is much more simple and rugged and one which has a considerably greater power output than a dc generator of comparable size.

Since in an alternator the field poles must rotate, a means of excitation voltage is applied to the field coils by a pair of sliprings on the rotor shaft. Sliprings do not present the problems that commutators do for two reasons: (1) The brushes through which current flows into the sliprings are in continuous contact with the rings, and therefore there is no occasion for sparking at the rings; (2) the voltage applied to the sliprings is seldom in excess of 250 volts.

The principles involved in winding the alternator armature are similar to those of the armature windings of dc generators. The coils forming the winding are lap-connected so that the output of the whole winding is the sum of the individual emf's induced into each of the coils composing the winding. Because of the greater efficiency and utility of three-phase alternators, most alternators are of the three-phase type. This means that the armature is composed of not one, but three separate windings. These windings are identical, their only difference being that they are spaced 120 electrical degrees apart. Alternators fall into three different construction categories, depending on the prime mover for which they are intended. There are the engine-driven alternator, the water-wheel alternator, and the turbine-driven alternator. Engine-driven and water-wheel alternators are relatively large low-speed machines, while turbine-driven alternators are relatively small high-speed machines. For their size, however, turbine-driven alternators have a considerably greater output than the other two types of generators.

Compared to the regulation of shunt generators, the regulation of alternators is poor. The thing which has greatest effect on the regulation of an alternator is the character of the load. When the load has a leading power factor, the terminal voltage of the alternator increases gradually as the load is increased. When the load has a unity or a lagging power factor, the terminal voltage falls as the load is increased--the rate of fall being greater for the lagging power factor. Since the load on most alternators generally has a lagging power factor, the terminal voltage would ordinarily vary considerably with changes in load. This does not happen because the output of alternators is usually regulated by a properly designed voltage regulator.

REVIEW QUESTIONS

1. In what two ways does an alternator differ from a dc generator?

2. Name the three types of alternators.

3. Why must an alternator be driven at a constant speed?

4. Compare the engine-driven alternator and the turbine-driven alternator with regard to type and number of poles used.

5. How many windings does the armature of an alternator usually have?

6. In what units is an alternator rated?

7. When the load is light, is it better to use one alternator operating at its rated capacity, or is it better to share the load between two alternators operating at less than their rated capacity?

8. Name some of the losses to which an alternator is subject.

ANSWERS TO REVIEW QUESTIONS

1. One way in which an alternator differs from a dc generator is that the field is rotated instead of the armature. Another difference is that the alternator is separately excited.

2. From the standpoint of construction, alternators fall into three classes--engine-driven, water-wheel, and turbine-driven alternators.

3. An alternator must be driven at a constant speed so that the frequency of the output voltage will be constant.

4. The rotor of an engine-driven alternator is large and consists of a large number of salient poles. The rotor of a turbine-driven alternator is a smaller smooth cylindrical steel casting containing only two poles.

5. Since most alternators have three phases, the armature has three windings 120° apart.

6. An alternator is rated in kva (kilovolt amperes).

7. Since an alternator is most efficient when it operates at its rated capacity, it is best when the load is light to use just one alternator.

8. Some of the losses to which an alternator is subject are field loss, friction and windage loss, core loss, and armature I^2R loss.

5. Alternating-Current Induction Motors

Most of the electrical power available in this country is of the alternating-current type rather than the direct-current type. It is not surprising, therefore, that many types of ac motors have been developed, each type being best suited to a particular application.

Alternating-current motors may be conveniently divided into two main classes; namely, induction motors and single-phase motors.

The use of ac motors is very widespread in both domestic and industrial applications. This is, no doubt, due to the greater prevalence of alternating-current electricity over direct-current electricity.

The alternating-current motors presented in this chapter are the induction motor, synchronous motor, universal motor, repulsion motor and shaded-pole motor. Of these, the most widely used ac motor is the induction motor.

The induction motor is very simple and rugged. Since it has neither commutator nor sliprings, its main parts are the stator and the rotor.

The spacing of the stator poles and the phasing of the currents which energize these poles are such as to generate a magnetic field that revolves around the rotor. The speed of revolution of the field is known as the synchronous speed and is related to the frequency of the applied voltage in cps. The revolving magnetic field induces currents in the rotor whose associated magnetic fields react with the revolving magnetic field in such a way as to produce torque.

While an induction motor may operate with one, two, or three phases of alternating current, only two- and three-phase motors are self-starting. (Split-phase motors are self-starting but this is because they operate as two-phase motors on starting.)

Most induction motors employ squirrel-cage rotors and, for this reason, are called squirrel-cage motors. Such motors have excellent operating characteristics but poor starting characteristics. In spite of the fact that their starting currents may be as much as six or seven times the rated value, their starting torque is relatively low. In order to prevent line disturbances on starting, most of the larger squirrel-cage motors have automatic starters or compensators which place resistance in series with the line on starting and gradually cut out the resistance as the motor approaches its operating speed.

The starting characteristics of induction motors can be considerably improved by replacing the squirrel-cage rotor with a wound rotor. This rotor has sliprings through which resistance can be introduced into the rotor. The introduction of resistance enables the motor to develop higher values of torque at greater values of slip. Thus, by placing enough resistance in series with the rotor, the motor can be made to have a very high starting torque. Then as the motor comes up to speed the rotor resistance can be cut out.

Resistance in a wound rotor not only gives the motor a high starting torque but also provides a means of controlling the speed of the motor. The greater the rotor resistance, the lower the speed of the motor. Even with all of the starting resistance cut out, the wound rotor has considerably more resistance than the squirrel-cage rotor. Thus, the high starting torque is obtained at the expense of reduced efficiency and poor speed regulation.

If an ac voltage is applied to the armature windings of an alternator and then the rotor is brought to synchonous speed, the rotor will continue to turn at synchronous speed.

Such a machine is called a synchronous motor. The usual way of starting a synchronous motor is to open the dc field and operate the motor as an induction motor until it reaches synchronous speed. At this time, the dc field is closed and the machine will continue to turn at synchronous speed. The fact that the motor turns only at synchronous speed makes it ideally suited for operating clocks and other timing mechanisms. Another advantage of synchronous motors, particularly the large three-phase type, is that by adjusting their dc field they may be made to act either inductively or capacitively. This is of importance to many industrial applications in which the ability to control power factor is essential to operating efficiency.

REVIEW QUESTIONS

1. Why can the induction motor be considered as a special type of transformer?

2. What conditions are necessary in a two-phase induction motor in order to develop a revolving field?

3. What is the synchronous speed of a 2-pole, 60-cycle, two-phase induction motor?

4. What happens to the synchronous speed of a motor as the number of poles of the motor is increased?

5. What is the revolution slip of an induction motor?

6. Compare the slip of a squirrel-cage motor with that of a wound-rotor induction motor.

7. What is the advantage of a squirrel-cage motor over a wound-rotor induction motor?

8. What two advantages does a wound-rotor induction motor have over a squirrel-cage motor?

9. Under what condition does a synchronous motor develop torque?

10. What is the effect of varying the excitation field of a synchronous motor?

ANSWERS TO REVIEW QUESTIONS

1. An induction motor can be considered a special type of transformer because the voltages in the rotor needed to produce torque are induced into the rotor by currents in the stator.

2. The conditions necessary to develop a revolving field in a two-phase induction motor are that the two currents be in both space and phase quadrature.

3. The speed of this two-phase, 2-pole induction motor is 60 rps.

4. As the number of poles of an induction motor is increased, its synchronous speed increases.

5. The revolution slip of an induction motor is the difference between the speed of revolution of the revolving field and the actual speed of rotation of the rotor.

6. A squirrel-cage motor has considerably less slip than a wound-rotor induction motor.

7. The squirrel-cage motor has higher efficiency and better speed regulation than the wound-rotor induction motor.

8. The wound-rotor induction motor has a higher starting torque than the squirrel-cage motor and has provisions for adjusting its speed.

9. A synchronous motor develops torque only when the rotor is turning at the synchronous speed.

10. By varying the excitation field of a synchronous motor, it is possible to adjust the power factor of the motor.

6. <u>Single-Phase AC Motors</u>

The electrical power generally available to homes, small businesses, and industry throughout this ountry is 60-cycle, single-phase, 110- and 220-volt alternating current. In view of this, an understanding of the operation of single-phase motors is of considerable importance.

A small series motor of conventional design will operate equally well with direct-current or alternating-current electricity. The reason for this is that the polarity of the applied voltage has no effect on the direction of rotation of the dc motor. Large series motors will not work well with alternating current unless they have been specially designed for operation with either type of electricity. Motors that operate equally well with either type of electrical current are known as universal motors.

An ordinary dc motor will operate with single-phase ac if its brushes are short-circuited and if the brush axis is at an angle of about 20° with respect to the axis of the main field. On passing through the armature, the current sets up a magnetic field about the armature which is in opposition to the magnetic field existing at each pole. As a consequence, a continuous force of repulsion exists between the main poles and the armature, which causes the armature to rotate continuously. A motor that operates in this manner is a repulsion motor. An advantage of this type of motor is that it has a high starting torque.

The shaded-pole motor is similar to the induction motor that the rotor currents required to produce rotation are induced into the rotor. It differs from the induction motor in that the magnetic axis, rather than revolving continuously about the rotor, merely shifts rhythmically from one side of each pole to the other. Shaded-pole motors have low starting torque, low efficiency, and little overload capacity.

Once started, a two-phase induction motor will continue to operate with reduced efficiency if one of the phases is removed. Then the motor is operating as a single-phase induction motor.

Since the single-phase induction motor is not self-starting, its stator is generally supplied with two sets of windings, the main winding and the auxiliary winding. The main winding usually has many turns and is highly inductive; the auxiliary winding is physically displaced with respect to the main winding by an angle of 90°, contains fewer turns, and generally has either a capacitor or resistor in series with it. The series combination of the auxiliary winding and capacitor or resistor, whichever the case, connects to the line through a centrifugally operated switch. On starting, the switch is closed and the phase difference between the two branch currents results in a reaction very similar to that existing in the conventional two-phase induction motor. When the motor comes up to speed, the centrifugal switch opens, and the motor continues to operate with current only in the main winding. For the purpose of making a distinction, single-phase induction motors containing resistance in series with the auxiliary winding are referred to as resistance-start motors, and those containing a capacitor in series with the auxiliary winding are referred to as capacitor-start motors. Capacitor-start motors have better operating characteristics than resistor-start motors and for this reason are much more widely used.

REVIEW QUESTIONS

1. What is a universal motor?

2. What determines the direction of rotation of a repulsion motor?

3. What is done to make a single-phase induction motor self-starting?

4. What is the advantage of a capacitor-start motor over a resistor-start motor?

ANSWERS TO REVIEW QUESTIONS

1. A universal motor is a series dc motor which will operate with either dc or ac electricity.

2. The direction of rotation of a repulsion motor depends on which side of the main field axis the brush axis is placed.

3. A single-phase motor is made self-starting by splitting the one phase into two phases. Then the motor starts as a two-phase motor.

4. A capacitor-start motor develops considerably more torque than the resistor-start motor and operates more efficiently.

Appendix 2

ENGINEERING REGISTRATION

Registration as a _professional engineer_ through a state board of registration is the only legal basis for the public practice of engineering. All state statutes and laws currently exempt engineers employed by industry or the Federal government; efforts are under consideration in some states to eliminate at least the industry exemption. The basis for engineering registration (licensing) is the obligation of the state for protection of the safety, health, and welfare of the public. Each state, territory, and the District of Columbia has a formal statute under which engineers are examined and registered; implementation of the mechanics of the registration process is carried out under "boards of registration." Registration to practice in any state must be obtained from that state; approval of an applicant registered in another state (comity) is based on satisfaction of the minimum standards of each board to which application is made. Each state is autonomous and therefore state statutes and board regulations are not uniform; the general range of requirements and procedures will be discussed later.

The practice of engineering is defined as based on an application of the basic principles of mathematics, physical sciences and the engineering sciences. Boards of registration evaluate education and experience as a basis for registration. While engineering is one of the slowest of the professions to formally require minimum educational requirements some states have or will shortly require an accredited engineering degree as the minimum standard for consideration and in almost all states registration is facilitated by the holding of a degree. Experience "of a nature satisfactory to the board" is a requirement by all boards. While some variation exists between boards in the amount of experience required, it, in all cases, must be of a professional nature (detailed definitions relating to experience will appear later).

While, as noted, each board of registration is autonomous, there is a general pattern to their procedures even though details and specific requirements may vary. The evaluation of a candidate by a board involves a judgment of education and experience. The judgment process may include formal examination in (1) fundamentals of engineering (often referred to as the Engineer-in-Training--EIT--examination), and (2) Principles and Practice of Engineering. Both examinations are eight hours in length and most states utilize common instruments prepared by the National Council of Engineering Examiners (a non-legal cooperative body established by the state boards to facilitate their operations; the NCEE also provides a "certification" service for engineers requiring registration in a number of states which enables state boards to simplify their evaluation processes.) Educational and experiential requirements for admission to examination vary--see Charts 1 and 2--and each individual must consult the board of registration in the state of residence for specific information.

Chart 1 summarizes the input processes, and the points outlined below are keyed to the chart. It must be emphasized again that the statutes/regulations of each board are unique as are interpretations within them. This is a broad and general overview only.

1. <u>Experience</u>. In some limited number of states, engineers of long practice and/or outstanding reputation (whose work has been in industry or through governmental agencies not requiring registration) may be admitted to candidacy for registration and to the P & P examination upon successful passage of which they will be registered. An age clause also exists in some states, satisfaction of which may allow direct registration based solely upon experience at this time. Comity normally is not available to such registrants.

2. <u>Experience</u>. States not having a formal educational requirement may allow an applicant having appropriate experience to take the Fundamentals Examination. Successful passage establishes EIT status, and after further experience, admission to the P & P examination is allowed. (A further permutation, also applicable to <u>3.</u>, <u>4.</u>, and <u>5.</u> below, in some limited number of states, allows registration after the EIT without examination after completion of a satisfactory period of professional experience.)

CHART 1

STATE REGISTRATION BOARDS
EVALUATION & ACTIONS

REGISTRATION AS A PROFESSIONAL ENGINEER

BOARD JUDGEMENT & DECISION

& AGE

PRINCIPLES & PRACTICES EXAMINATION

EXPERIENCE

(see parenthetical explanation, 3.)

FUNDAMENTALS EXAMINATION (E.I.T.)

QUALIFICATION OF APPLICANT*

1. EXPERIENCE

2. EXPERIENCE

3. ENGINEERING DEGREE (ECPP ACCREDITED)

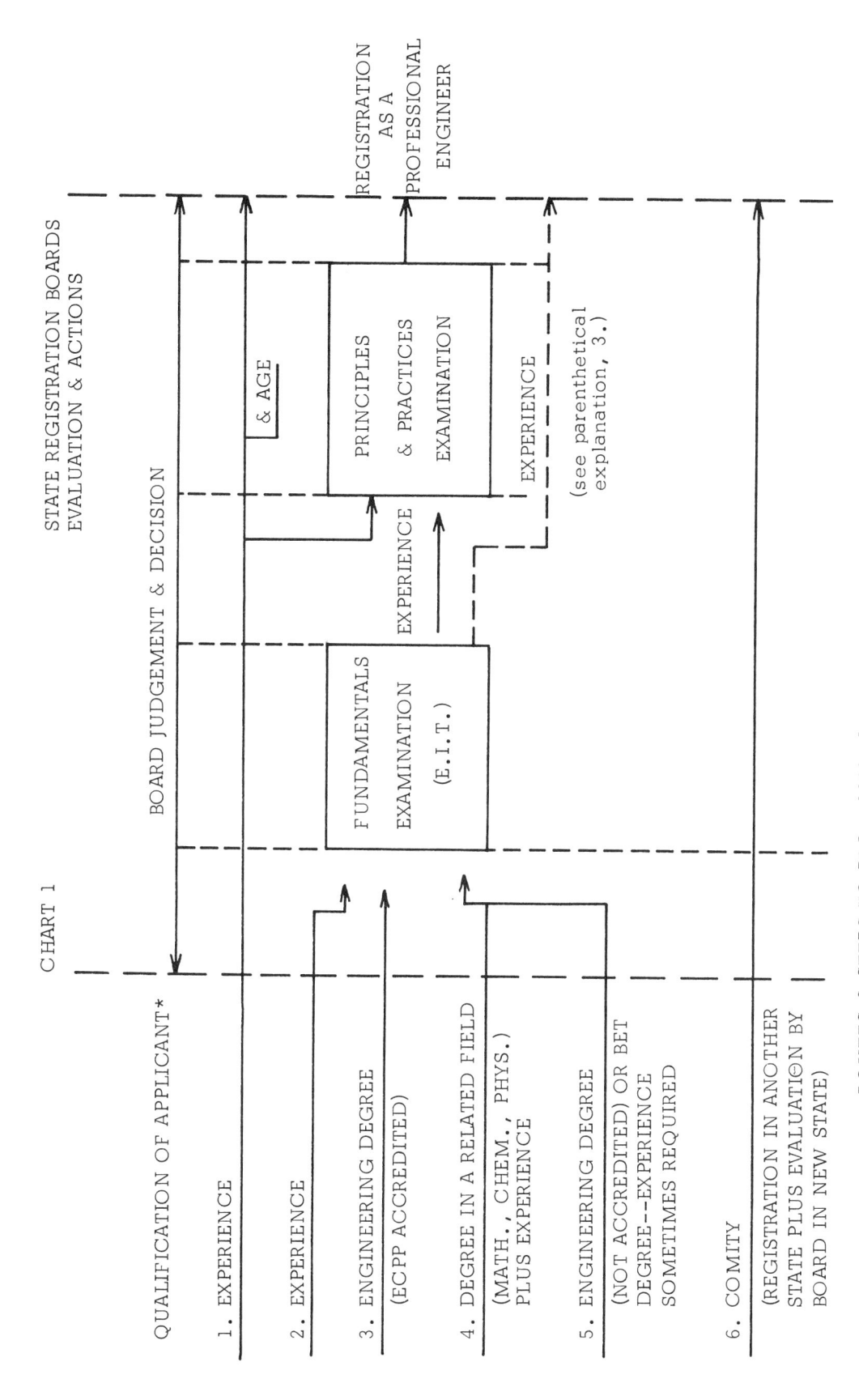

4. DEGREE IN A RELATED FIELD (MATH., CHEM., PHYS.) PLUS EXPERIENCE

5. ENGINEERING DEGREE (NOT ACCREDITED) OR BET DEGREE--EXPERIENCE SOMETIMES REQUIRED

6. COMITY (REGISTRATION IN ANOTHER STATE PLUS EVALUATION BY BOARD IN NEW STATE)

ROUTES & STEPS TO PROFESSIONAL ENGINEERING REGISTRATION

*See attached summary for explanation & expansion.

```
                    ┌─────────────────┐
                    │   PRINCIPLES &  │
                    │   PRACTICE      │
                    │   EXAMINATION   │
                    └─────────────────┘
                         ↑    ↑
   ┌─────────────────┐   │    │
   │  FUNDAMENTALS   │   │    │
   │    (E.I.T.)     │───┘    │
   │  EXAMINATION    │        │
   └─────────────────┘        │
         ↑    ↑               │
         │    │         EXPERIENCE (1)
                        ─────────────
FORMAL EDUCATION--CHART 2    supplemented by
                             "preparatory" or
(3)*, (4), (5)--sometimes    "refresher" courses
with "refresher courses"

EXPERIENCE (2)--
─────────────
supplemented by "preparatory"
or refresher courses
```

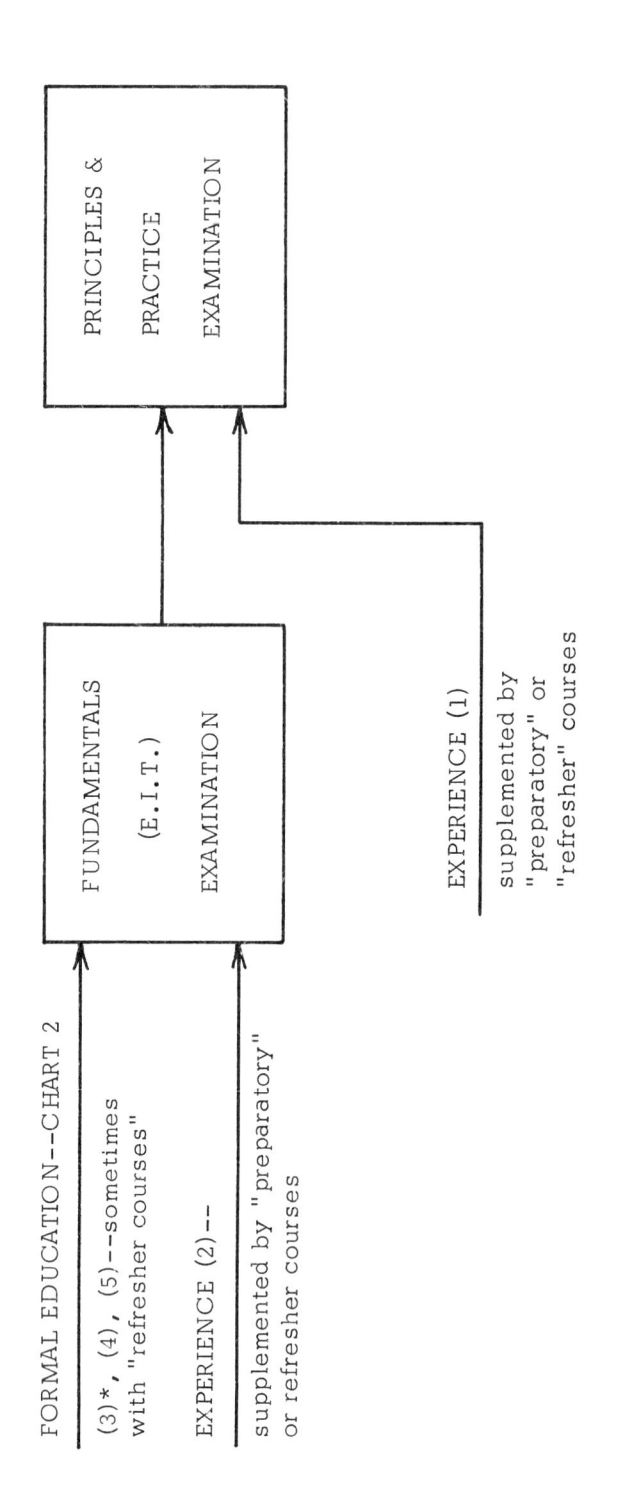

EDUCATIONAL PREPARATION FOR REGISTRATION EXAMINATIONS

---
*See Chart 1.

ENGINEERING REGISTRATION / 271

3. <u>Engineering Degree</u> (ECPD accredited). Such a graduate is regularly admitted to the Fundamentals Examination and upon successful completion is granted EIT status. (It may be of interest to note that the ECPD accredited graduates who pass this examination may range from 60 to 90% only.)

Chart 2 is in turn keyed to Chart 1, and expands on the educational preparations for both the fundamentals and Principles and Practice Examinations.

Experience requirements in time are variable but definitions of engineering experience are reasonably standardized (although individual board interpretations may vary). The following is quoted from a typical application form from a state board:

1. <u>Sub-Professional Work</u> is to cover time spent as Rodman, Chairman, Instrumentman, Inspector, Recorder, Draftsman, Computer, Tester, Superintendent of Construction, Clerk of the Works, Junior Engineer, or similar work; that is, positions in which the responsibility is slight and the individual performance of a test, set and supervised by a superior, is all that is required. It shall also include the time during which he has been occupied in engineering before the applicant is 21 years of age, except as modified by the statement in regard to education in the definition of Professional Work.

2. <u>Professional Work</u> shall include the time after the applicant is 21 years old, during which he has been occupied in engineering or land surveying work of a higher grade and responsibility than that above defined as Sub-Professional Work. If applying for registration under other than "graduation plus experience" list each academic year completed in an engineering school approved by the Board as 1 year of Professional Work, and list each academic year completed in a course other than engineering in a college or university of recognized standing as 1/2 year of Professional Work. Not more than 4 years of Professional Work shall be credited for undergraduate educational qualifications. List graduate study in engineering as Professional Work, but such study will not be credited for undergraduate educational qualifications. List graduate

study in engineering as Professional Work, but such study will not be credited as more than 1 year of Professional Work. Time spent in teaching of engineering subsequent to graduation shall be listed as Professional Work. The mere execution, as a contractor, of work designed by an engineer, or the mere supervision of construction of such work as foreman or superintendent, shall not be deemed to be professional work.

3. <u>Responsible Charge of Work Means:</u>

   a. In the field, the applicant must have had the direction of work, the successful accomplishment of which rested upon him, where he had to decide questions of methods of execution and suitability of materials, without relying upon advice of instructions from the superiors, and of supplying deficiencies in plans or correcting errors in design without first referring them to higher authority for approval, except in cases where such approval is a mere matter of form.

   b. In the office, the applicant must have had to undertake investigations, or carry out important assignments, demanding resourcefulness and originality, or to make plans, write specifications and direct drafting and computation for designs of engineering with only rough sketches, general information and field measurements for reference and guidance.

   c. In teaching, the applicant must have taught in engineering schools of recognized standing, and must had had, at least, a grade of assistant professor, or its equivalent.

4. <u>Design</u> means all that is given above as responsible charge in the office and more. One qualified to design must be able, in the case of any desired piece of engineering, to meet the exigencies of the case, to fulfill the requirements of local circumstances and conditions, and yet not violate any of the canons of engineering. His plan, when executed, must successfully answer the purpose for which it was desired.

As noted in the previous material each state registration board is autonomous, and individuals planning to apply for EIT status or professional registration should consult the board in his/her state of residence for specific information and application forms. However, there are a number of procedures which are essentially common. These will be discussed here.

1. Engineer-in-Training Status. The application form and procedure for EIT status is in most states different and simpler than for professional registration. If the applicant holds an engineering degree from an accredited institution, no experience record is required and only certification of such graduation and references, usually three to five from faculty members, are requested. Applicants holding a non-accredited engineering degree, a bachelor of engineering technology, or a degree in an associated field (physics, chemistry, mathematics, etc.) may be required to offer evidence of experience of an engineering nature ranging from one to four years. This experience must be documented in detail as to type, extent and level, and normally supported by a certification statement from an immediate supervisor during each time period. Experience of a non-engineering related nature will not be accepted. With a small number of exceptions states will accept for consideration applications from individuals not holding a degree, but who have an adequate amount of engineering-related experience or a combination of education and experience. Adequacy in time ranges up to twelve years, although four to six years is more the norm. As for the non-accredited degree applicant the education and experience must be of a clearly engineering-related nature. Work as a technician in support of engineers, if of a high technical level, is normally acceptable.

After the board has evaluated and approved the application, the candidate is informed of the time and place at which the Fundamentals of Engineering (EIT) Examination will be offered. The current form of the examination used by most states (a standard instrument prepared through the National Council of

Engineering Examiners) involves a total of eight hours divided into two four-hour periods. The examination is usually open book--i.e., printed reference materials and texts may be used. Some boards, however, limit the number of volumes which may be brought to the examination. The content of the examination includes:

a. Mathematics
b. Chemistry
c. Statics
d. Dynamics
e. Mechanics of Materials
f. Fluid Mechanics
g. Thermodynamics
h. Electrical Theory
i. Economic Analysis

For individuals who have been away from formal study for some time, several aids have been available to assist them in preparation for the examination. These include copies of earlier examinations* and solutions from the National Council of Engineering Examiners, self-study courses from several proprietarial sources, and "refresher" courses offered by universities and colleges. These aids may be useful, but care should be taken to avoid undue reliance on them; a solid grasp of the fundamentals of the areas covered is the only way of assuring successful passage. The practice examination in this publication reflect the most recent EIT Examination.

2. <u>Professional Registration</u>. Professional registration is granted by boards only after a vigorous and careful evaluation of the candidates qualifications. As noted earlier EIT status or the passing of the Fundamentals Examination is the most common basic qualification, followed in turn by a minimum of four years of engineering experience. The range and type of experience

---

*NOTE: Prior to 1971, the form of the examination has been changed considerably, but the content is similar.

was outlined earlier, and applicants must be prepared to state specifics of each experience, including a breakdown by decimal tenths of years for each category and the name of an immediate supervisor who can attest to the accuracy of each listing. In addition, four or more registered professional engineering personal references are normally required. Except where experience is long and of a notably high level or outstanding quality, a general requirement also is the successful passage of an eight-hour examination in the Principles and Practices of Engineering in a specific professional field. As with the EIT examination several preparatory aids are available to those planning to take the examination. These again include access to prior examination, self-study materials, and refresher courses. Since the P & P examination is of a narrower and more specialized nature, less emphasis on fundamentals may be expected, but the candidate must be sure that his experience in a specific field is sufficiently broad to cover the whole of the specialty chosen.

Common and critically important to both the EIT and professional registration processes is the analysis of and presentation of the applicant's engineering experience. It is strongly recommended that a conservative approach be taken in stating both level and length of time for all activity.

STATE BOARD OF ARCHITECTURAL AND ENGINEERING EXAMINERS

AND

BOARD FOR EXAMINERS IN LANDSCAPE ARCHITECTURE

1. GENERAL INFORMATION

Date.. , 19.......

I am applying for registration as a Professional Engineer..

Name in full..

Residence
Address .... City............................ State ....

Present
Position ....Company.....................................

Business
Address ....City............................ State............

Birthplace .... Date ....Citizenship ....

> Attach securely in this space a recognizable unmounted picture taken within 90 days of submission of application. Face should not be less than ¾" wide.
> Sign and place date that picture was taken across bottom of picture.

If foreign born, when did you come to U. S.? ....

Indicate what citizenship papers have been taken out...

Are you engaged at present in the practice of engineering?......................................

If not, what is your occupation? ....

Name, as you wish it to appear on your seal ....

In what branch or branches of engineering do you consider yourself best qualified to

practice? ....

2. REGISTRATION

Are you registered?.................. State..Number........................Category...

Was written or oral examination taken?.................. How many hours of EIT?................................PE ....

Is Certificate now in force?................If not, why?..

If you are registered in more than one State, name State of first registration ....

Name other States in which you are registered ....

Has any State denied registration to you or revoked it?..Give details separately............................

3. REFERENCES

Give names and addresses of five engineers who are not relatives, not members of the   nor those whose names appear on the inside of the sheet under 5, Experience Record. These references should be able to vouch for you on the basis of personal knowledge, first as to your character and reputation and second as to your engineering experience and ability. Three of your references must be registered professional engineers.

Name	Address	Business Relation to Applicant	Has Known Applicant Since

FORM NO. 3

4. EDUCATION

State in chronological order, the name and location of each high or preparatory school, college, university, or technical school attended, the time spent at each, and if graduated, the year of graduation. Also list graduate work, evening schools attended, correspondence courses taken, and academic research work.

NAME AND LOCATION OF EDUCATIONAL INSTITUTION Indicate evening school or any other special type.	YEARS From-To	DATE GRADUATED	TECHNICAL COURSE	DEGREE RECEIVED

5. EXPERIENCE RECORD

"Other work time" includes all work which is not Engineering.
Each of the five columns under "Engineering Time" must be filled out for each engagement. Use zeros where necessary, but do not leave blank spaces, and do not use the word "yes".
The time in "Sub-Professional Work" plus the time in "Professional Work" must equal the time entered under "Total Time."
Any portion of the time entered in column (2) as "Professional Work" which has been in "Responsible Charge" should also be entered in column (4), and any portion which has been in "Design" should also be entered in column (5); columns (4) and (5) may overlap relative to time since each is independent.

Engagement Letter (a, b, etc.)	DATE From — To	TITLE OF POSITION, NAME OF EMPLOYER, AND CHARACTER OF EACH ENGAGEMENT Make statement brief and concise, designating each engagement or change in position by a separate letter; include magnitude and complexity of work on which engaged, your duties and degree of responsibility; any necessary amplifications may be made on separate sheet. Start with first position as engagement "a".	OTHER WORK TIME	ENGINEERING TIME (Years in Decimals to Tenths)					NAME, TITLE, & ADDRESS of some person (not deceased) familiar with each engagement, preferably the person to whom applicant reported.
				(1) Sub-Professional Work	(2) Professional Work	(3) Total Time (1) plus (2)	(4) Responsible Charge	(5) Design	

SUMMARY BY APPLICANT

SUMMARY AS VERIFIED BY BOARD—Not to be filled in by applicant

6. MEMBERSHIP IN SOCIETIES, ASSOCIATIONS, Etc. (Professional and Scientific)

| Name of Organization | Address | Grade of Membership | Date Joined |

INCLUDE INFORMATION ABOUT PROFESSIONAL ACTIVITIES, PUBLICATIONS, PATENTS, Etc.

7. CODE OF ETHICS

I acknowledge the receipt of a copy of the "Code of Ethics" as promulgated in accordance with   and adopted by the

By my signature hereto, I state that I have studied this Code and that I will endeavor to guide my conduct in accordance with its principles to protect all aspects of the public interest and to uphold the character of the Engineering Profession.

Furthermore, I will co-operate with the   by reporting to them all violations of the Registration Law which come to my attention.

In making this pledge, I recognize that the practice of a profession carries specific individual responsibilities which I intend to fulfill.

Date...Signed..

8. AFFIDAVIT

STATE OF ....

County of .... } ss.

...., being first duly sworn, deposes and says: I, the Applicant named in this application, have executed the contents hereof, and to the best of my knowledge and belief the foregoing statements are true in substance and effect and are made in good faith, with no information being suppressed which might affect this application.

Subscribed and sworn to before me this

.................................day of..., 19......... ....
(Signature of Applicant)

(SEAL)

My Commission expires ....
(Signature of Notary Public)

9. RECORD OF COUNCIL
(This space not to be used by Applicant)

Name of Applicant..Date Application received.................................

Amount of fee paid..Considered by Council ....

Examination given......................................Date ....Rating..

Action of Council ....

Certificate Issued.. Number..

Secretary's Notes:

**HOW TO APPLY TO REGISTRATION BOARDS
FOR ENGINEER-IN-TRAINING STATUS**

Each state board is autonomous and functions under its own statute and regulations. Detailed information on procedures and requirements can only be obtained directly from the registration board in the state of application. The addresses of State Registration Boards follow this section. However, there are some general and common components to these requirements which are summarized in this section.

Each registration board deals with the educational and experience record of the applicant as presented in a formal application and as verified by the references provided. The evaluation process carried out by a board is thorough and exacting. While all board members are highly competent professional engineers, it must be recognized that because of the broad and diverse spectrum of education and types of experience associated with engineering, they may not be familiar with some less conventional areas. Specialized educational and technical experiences may well fall into this catagory, and considerable detail will probably be required in any application. In most states, the EIT application is not complex, but particularly where experience is offered in lieu of, or partial substitution for, a degree, boards will normally require significant detail on experience and the latter section on this will generally encompass both types of applications.

1. Presentation of Education. The optimum (and increasingly expected) educational base for both EIT status and professional engineering registration is graduation from an ECPD-accredited engineering curriculum. The holding of such a degree assures at least a minimum educational background encompassing the mathematical and physical sciences, the engineering sciences, engineering analysis and design, and, the humanities and social sciences (that almost all boards also require an examination in engineering fundamentals--i.e., the EIT examination--and that not all accredited curriculum graduates suc-

cessfully pass it suggests that formal education is not in itself necessarily complete or wholly satisfactory). Boards will, however, generally examine other education or experience in lieu of education in the light of the scope and depth of an accredited curriculum.

Other educational backgrounds presented for consideration may range from a nonaccredited engineering degree through a baccalaureate degree in mathematics or science or in engineering technology to the partial completion of a program in any of these areas. Specialized professional/technical training may be offered for consideration. A Board will, normally, then evaluate the education presented in terms of areas covered and depth with particular emphasis on the level of basic mathematics--normally through calculus and differential equations--utilized. Conventional curricula and courses may be described using the name of the program and the degree and/ of the portion completed. Non-regular courses or programs should be described in detail including topical areas covered, and particularly noting the level and type of mathematics and/ or science base required and used. Where available, certificates of completion and course outlines should be included (catalog descriptions may be sufficient, but are generally, except for standard courses, best supplemented by additional detail). Most applicants will have some formal educational background to present, but some board regulations allow for supplementing or substituting for education appropriate and acceptable engineering experience. The form and nature of this experience is outlined in the next section.

An increasing number of states have (or will soon) established graduation with an ECPD accredited degree as a minimum requirement for the EIT Examination.

An applicant for EIT status normally must have either the basic formal education, or a combination of education and/or experience in lieu of part or all of the educational requirements. Following satisfaction of the board of this requirement, the applicant will be informed by the board of the time and place at which the EIT examination will be given. (Information on the general content of and preparation procedures for the EIT Examination are covered in a separate section.)

2. <u>Presentation of Experience</u>. Most statutes and regulations assign to the board full responsibility and authority to establish minimum periods and content of engineering experience—i.e., "experience of a nature satisfactory to the board." Two or more years of engineering experience is usually required for EIT eligibility. Again, while specifics may vary, the general expectations are as outlined in previous sections, <u>Professional Engineering Registration</u>. In brief, summary of engineering experience must be of a professional nature, which in addition, to the specific function and level of responsibility builds upon the application of the fundamental principles of mathematics, the basic physical sciences, and the engineering sciences in the solution of problems.

Each experience must be described in detail as required on the application form for the particular state, but is normally divided into semi-professional and professional components with, in turn, the professional portion allocated to "general professional," "responsible charge," and "design" by decimal tenths of years, the totals of which must satisfy the minimum board requirements for each category. The name, title, and contact address of the immediate supervisor for each period is normally required so that the board may obtain verification as desired. It is particularly important for experience areas not familiar to the board that reasonable detail be provided for the board's information. If adequate space is not available on the form, appendices keyed to each experience block should be attached spelling out the detailed duties and responsibilities as related to the experience expectations previously outlined.

Experience is considered in the context of "professional" and in the words of most statutes "of a nature satisfactory to the Board." Definitions of satisfactory professional experience may be presented in either an exclusionary or inclusionary way. Typically excluded is technical work of a "routine nature," in "minor positions," "of a routine nature in which the task performed is set by detailed instructions," "under close supervision," or "which requires or allows little or no individual initiative or responsibility." Included in allowable and acceptable work is effort involving a knowledge and individually responsible application of mathematics, physical sciences and engineering principles to the solution of technical problems; general and broad professional supervision is appropriate but primarily in an overview or review sense. In addition, "responsible charge" during a certain portion of these efforts is normally expected; this basically is carrying **out of** the above kind of effort in an environment in which successful accomplishment is rested basically upon the individual and little or no advice or counsel from superiors or supervisors was available or expected and normally is essentially a matter of form.

It is important to distinguish between the minimum period of "lawful practice in professional engineering work" in contrast to "experience in engineering work."

The law in some states thus indicates a difference between "engineering work" and "professional engineering work," and between "experience" and "lawful practice." The essential difference between the contrasting terms is one of responsibility for the control and direction of the investigation, design, construction or operation of engineering work requiring initiative, professional skill, and independent judgment. <u>Experience</u> in engineering work may be of a lower degree of responsibility, such as work done under direction or supervision. It may include the period when the younger engineer is obtaining his training in practice. Lawful practice in <u>professional</u> engineering work occurs when one exercises authority and responsibility in his engagements equivalent to that which is assumed by a professional engineer holding himself out in relation of agent to client.

Typically, if a Board allows registration without examination, it requires at least twelve years of lawful practice in professional engineering work, of which five years or more must have been in responsible charge of important engineering work, and a minimum age ranging from 35 to 45 years. But, first the applicant must <u>submit evidence</u> that he is an "engineer of recognized and established standing in the engineering profession." Regardless of education and experience, such recognition is essential to qualification under this category. What constitutes such evidence in addition to significant professional/technical accomplishment?

Testimony from one's employers or associates contributes to this "evidence of recognized and established standing in the engineering profession," but in itself is not sufficient. The engineering profession has many ways of offering such recognition. For example, invitations to take part in the activity of professional committees dealing with engineering standards, invitations to present or discuss technical papers presented before technical societies, invitations to review articles presented for publication in technical magazines, invitations to prepare technical articles for publication, membership in those technical societies known to be selective regarding membership qualifications, membership on national or international technical committees, the acceptance of record of one's own engineering work in the form of publications or patents, selection for agency or departmental technical committees or review boards, election to office in one of the professional associations or societies, and many similar activities are all explicit evidence of recognition by the engineering profession.

A board does not require that any one individual take part in all these activities, but a reasonable participation in some of them is regarded as most desirable. Many other factors, including educational background and records of specific engineering projects which clearly show the applicant's participation and responsibility figure in the Board's deliberations and contribute to the evaluation of the applicant's recognized standing in the profession. It should be noted, however, that registration without examination (by eminence) is relatively rarely allowed. Particularly in the absence of the applicant's holding a formal engineering degree, registration boards will require satisfactory passing of the EIT Examination and Principles & Practice Examination even though the experience record may be satisfactory.

Responsible Charge

In the field, the applicant must have had the direction of work, the successful accomplishment of which rested upon him, where he had to decide questions of method of execution and suitability of materials, without relying upon advice or instruction from his superior, and of correcting deficiencies in plans or errors in design without first referring them to higher authority for approval, except where such approval is a mere matter of form.

In the office, the applicant must have undertaken investigations, or carried out important assignments that demand resourcefulness and originality, or made plans, written specifications and directed the drafting and computations made in connection with engineering work when guided solely by rough sketches, general information and field measurements.

Engagements with the same employer are to be determined by changes in kind of work or in degree of personal responsibility. The exact status of the applicant in respect to supervision must be made clear; this can be done briefly by reference to an organization diagram indicating the flow of responsibility. Work should be classed as sub-professional whenever it does not utilize a knowledge of mathematics, the physical sciences, and principles of engineering, in an original manner; that is, when the work is of a routine character according to established standards of procedure; also, when technical work is done under supervision by direction of method to be followed, i.e., when the initiative and responsibility are elsewhere.

In engineering teaching, the applicant must have taught in an engineering school of recognized reputation, and must have had, at least, the title or position of associate professor, or its equivalent.

Design

All that is given above as responsible charge of work in the office, and more, is involved. One qualified to design must be able, in the case of any desired piece of engineering, to meet the exigencies of the case, to fulfill the requirements of local circumstances and conditions, and yet not violate the canons of engineering. His plan, when executed, must successfully answer the purpose for which it was intended.

On $8\frac{1}{2}$ X 11 inch sheets, the technical detail of each engagement should be set forth, the engagement being numbered as in the application form. The mathematics, and principles of physical sciences and engineering, which were used by the applicant in each engagement should be stated specifically. The technical problems which were encountered, the method of investigations, and the procedure in design and economic analysis, should be described. These statements constitute the bulk of the evidence which the applicant must submit in his application, so it is incumbent upon the applicant, that he reveal exactly how he utilized in practical experience, the knowledge required by the law.

In each case a Board is required to evaluate the character of an applicant's experience to determine if it has been satisfactory and adequate to indicate competency to practice. The description of experience should, therefore, be so prepared as to permit these determinations.

In evaluating engineering experiences for consideration by a state board of registration, the following should be addressed:

1. A specific description of the function performed and over what period of time and including,

2. Its sophistication and/or complexity,

3. The degree of routineness vs. individual judgment and initiative called for,

4. The amount and level, breadth, and scope of education and/or training requested,

5. The amount of individual responsibility and authority vs. supervision received or required,

6. And the level and amount of supervisory authority and responsibility exercised.

Engineering experience is work based on utilization of basic mathematical, scientific, and engineering principles and preparation, and carried out in a non-routine independently responsible manner.

Appendix 3

ADDRESSES OF STATE REGISTRATION BOARDS

The following is a list of all State Boards of Registration or Examiners for Professional Engineers. All of these State Boards or Departments are Member Boards of the National Council of State Boards of Engineering Examiners. Federal employees seeking professional registration or engineer-in-training status should write directly to the appropriate Board.

Professional engineering examinations, including the Fundamentals (EIT Examinations) and Principles and Practice of Engineering Examinations, are administered by the individual State Board, and application must be made to take the above examinations.

ALABAMA: State Board of Registration for Professional Engineers and Land Surveyors, 64 North Union Street, Room 607 Administrative Building, Montgomery, 36104. Telephone: Area code 205, 269-6208 or 269-6209. Miss Sarah E. Hines, Executive Secretary.

ALASKA: State Board of Architects, Engineers and Land Surveyors, Pouch D, Juneau, 99801. Telephone: Area code 907, 586-1501. Ms. Jean Boone, Executive Secretary.

ARIZONA: State Board of Technical Registration, 1645 W. Jefferson Street, Phoenix, 85007. Telephone: Area code 602, 271-4053. Edward G. Drewry, Executive Director.

ARKANSAS: State Board of Registration for Professional Engineers and Land Surveyors. PO Box 2541, North Little Rock, 72203. Telephone: Area code 501, 371-2517. Leland B. Bartlett, Secretary-Treasurer.

CALIFORNIA: State Board of Registration for Professional Engineers, 1021 O Street, Room A-102, Sacramento, 95814. Telephone: Area code 916, 445-5544. Don H. Nance, P.E., Executive Secretary.

CANAL ZONE: Board of Registration for Architects and Professional Engineers, Box 223, Balboa Heights, Canal Zone. Telephone: Balboa 52-7927. Barbara D. Peterson, Executive Secretary.

COLORADO: State Board of Registration for Professional Engineers and Land Surveyors, Room 102 State Services Building, 1525 Sherman Street, Denver, 80203. Telephone: Area code 303, 892-2396. Henry J. Ochs, Jr., Executive Secretary.

CONNECTICUT: State Board of Registration for Professional Engineers and Land Surveyors, Room 533, State Office Building, Hartford, 06115. Telephone: Area code 203, 566-3386. Ernest B. Gardow, Secretary.

DELAWARE: Delaware Association of Professional Engineers, 1508 Pennsylvania Avenue, Wilmington, 19806. Telephone: Area code 302, 656-7311. Ms. Ruth F. Letherbury, Executive Office Secretary.

State Board of Registration for Professional Land Surveyors, 1228 North Scott Street, Wilmington, 19806. Telephone: Area code 302, 658-9251 Ext. 475. Ms. Carolyn M. Rust, Executive Office Secretary.

DISTRICT OF COLUMBIA: Board of Registration for Professional Engineers, 614 H Street, N.W., Rm. 109, Washington, 20001. Telephone: Area code 202, 629, 4543. Henry B. Peterson, Secretary.

FLORIDA: State Board of Professional Engineers and Land Surveyors, Suite 100, 6990 Lake Ellenor Drive, Orlando, 32809. Telephone: Area code 305, 855-5970. J.Y.Read, Executive Director.

GEORGIA: State Board of Registration for Professional Engineers and Land Surveyors, 166 Pryor Street, S.W., Atlanta, 30303. Telephone: Area code 404, 656-3926. C.L.Clifton, Joint Secretary.

GUAM: Board of Engineering and Architectural Examiners, Dept. of Public Works, Government of Guam, Agana, 96910. Telephone No. 746-5831, Ext. 10. William A. McAlister, A.I.A., Secretary.

HAWAII: State Board of Registration of Professional Engineers, Architects, Surveyors and Landscape Architects, P.O. Box 3469, Honolulu, 96801. Telephone: Area code 808, 548-3086. Morris M. Comer, Executive Secretary.

IDAHO: State Board of Engineering Examiners, 110 North 27th Street, Boise, 83706. Telephone: Area code 208, 345-0735. Elton R. Leitner, P.E., Executive Secretary.

ILLINOIS: Department of Registration and Education, 628 E. Adams Street, Springfield, 62786.

Professional Engineers' Examining Committee: Mrs. Wilma Hofferkamp, Supervisor. Telephone: Area code 217, 525-4569.

Structural Engineers' Examining Committee: Frank Klein, Secretary. Telephone: Area code 217, 525-4744.

INDIANA: State Board of Registration for Professional Engineers and Land Surveyors, 1021 State Office Building, 100 N. Senate Avenue, Indianapolis, 46204. Telephone: Area code 317, 633-6790. Bernard T. Loeffler, Secretary.

IOWA: State Board of Engineering Examiners, State Capitol, Des Moines, 50310. Telephone: Area code 515, 281-5602. Mrs. Bernadine Millslagle, Assistant Secretary.

KANSAS: State Board of Engineering Examiners, 535 Kansas Avenue, Room 1202, Topeka, 66603. Telephone: Area code 913, 296-3288. Ellsworth E. Crowley, Executive Secretary.

KENTUCKY: State Board of Registration for Professional Engineers and Land Surveyors, University Station, Box 5075, Lexington 40506. Telephone: Area code 606, 255-5835.

LOUISIANA: State Board of Registration for Professional Engineers and Land Surveyors, 1055 St. Charles Avenue, Suite 415, New Orleans, 70130. Telephone: Area code 504, 581-7938. Daniel H. Vliet, P.E., Executive Secretary.

MAINE: State Board of Registration for Professional Engineers, State House, Augusta, 04330. Telephone: Area code 207, 289-3236. Sylvester L. Poor, Secretary.

MARYLAND: State Board of Registration for Professional Engineers and Land Surveyors, 1 South Calvert Street, 8th Floor, Baltimore, 21202. Telephone: Area code 301, 383-2447. Mrs. Hazel B. Wilkerson, Administrative Secretary.

MASSACHUSETTS: Board of Registration of Professional Engineers and of Land Surveyors, Room 1512 State Office Building, 100 Cambridge Street, Boston, 02202. Telephone: Area code 617, 727-3088. Samuel Valencia, P.E., Secretary.

MICHIGAN: Dept. of Licensing and Regulation, 1116 St. Washington Avenue, Lansing, 48926. Telephone: Area code 313, 373-3880. Board of Registration for Professional Engineers, Board of Registration for Land Surveyors. Jack C. Sharpe, Administrative Secretary.

MINNESOTA: State Board of Registration for Architects, Engineers and Land Surveyors, 1512 Pioneer Bldg., St. Paul, 55101. Telephone: Area code 612, 296-2388. Lowell E. Torseth, Executive Secretary.

MISSISSIPPI: State Board of Registration for Professional Engineers and Land Surveyors, PO Box 3, Jackson, 39205. Telephone: Area code 601, 354-7241. O.B.Curtis, Sr., P.E., Secretary. Mrs.Joyce V. Miller, Assistant Secretary.

MISSOURI: State Board of Registration for Architects, Professional Engineers and Land Surveyors, PO Box 184, Jefferson City, 65101. Telephone: Area code 314, 751-2678. Mrs. Olean Barton, Secretary.

MONTANA: State Board of Registration for Professional Engineers and Land Surveyors, LaLonde Building, Helena, 59601. Telephone: Area code 406, 449-3737. Mrs. Bernice Luck, Assistant Secretary.

NEBRASKA: State Board of Examiners for Professional Engineers and Architects, 512 Terminal Building, 941 "O" Street, Lincoln, 68508. Telephone: Area code 402, 471-2021. Arthur Duerschner, Executive Director.

NEVADA: State Board of Registered Professional Engineers, PO Box 5208, Reno, 89503. Telephone: Area code 702, 329-1955. John N. Butler, Executive Secretary.

NEW HAMPSHIRE: State Board of Registration for Professional Engineers, c/o Secretary of State, State House, Concord, 03301. Telephone: Area code 603, 224-4740. Stanley P. Sawyer, P.E., Secretary.

NEW JERSEY: State Board of Professional Engineers and Land Surveyors, 1100 Raymond Boulevard, Newark, 07102. Telephone: Area code 201, 648-2661. Carl E. Kastner, P.E., L.S., Secretary-Director.

NEW MEXICO: State Board of Registration for Professional Engineers and Land Surveyors, PO Box 4847, Santa Fe, 87501. Telephone: Area code 505, 827-2241. Ms. Gloria M. Garcia, Administrator.

NEW YORK: State Board of Examiners of Professional Engineers and Land Surveyors, New York State Dept. of Education, 99 Washington Ave., Albany, 12210. Telephone: Area code 518, 474-3846. Lawrence J. Hollander, P.E., Secretary.

NORTH CAROLINA: State Board of Registration for Professional Engineers and Land Surveyors, 1307 Glenwood Ave., Suite 152, Raleigh, 27605. Telephone: Area code 919, 833-0416. Col. Bolick A. Saholsky, Executive Director.

NORTH DAKOTA: State Board of Registration for Professional Engineers and Land Surveyors, PO Box 1264, Minot, 58701. Telephone: Area code 701, 839-5220. A.L. Bavone, P.E., Secretary.

294 / STATE REGISTRATION BOARDS

OHIO: State Board of Registration for Professional Engineers and Surveyors, 21 West Broad St., Columbus, 43215. Telephone: Area code 614, 469-3650. J. Roger Hayden, P.E., Executive Secretary.

OKLAHOMA: State Board of Registration for Professional Engineers and Land Surveyors, 401 United Founders Tower, Oklahoma City, 73112. Telephone: Area code 405, 848-5519. Mrs. Norma Unruh, Executive Secretary.

OREGON: State Board of Engineering Examiners, State of Oregon Dept. of Commerce, 325 13th Street, NE, Salem, 97310. Telephone: Area code 503, 378-4180, Ext. 1265. Paul D. Christerson, P.E., Executive Secretary.

PENNSYLVANIA: State Registration Board for Professional Engineers, Room 404, 279 Boas, Box 2649, Harrisburg, 17120. Telephone: Area code 717, 787-3151. Louise J. Franklin, Secretary.

PUERTO RICO: Board of Examiners of Engineers, Architects and Surveyors, PO Box 261, San Juan, Puerto Rico, 00904. Telephone: Area code 809, 724-2387. Justino Valles, Administrative Officer.

RHODE ISLAND: State Board of Registration for Professional Engineers and Land Surveyors, CIC Building, Room 17-A, 3rd Floor, 289 Promenade Street, Providence, 02903. Telephone: Area code 401, 277-2565. Mrs. Alice B. Gallenshaw, Assistant Secretary.

SOUTH CAROLINA: State Board of Engineering Examiners, 710 Palmetto, State Life Bldg., Columbia, 29201. Telephone: Area code 803, 787-5995. Mrs. Mary M. Law, Executive Secretary.

SOUTH DAKOTA: State Board of Engineering and Architectural Examiners, 2040 West Main Street, Suite 212, Rapid City, 57701. Telephone: Area code 605, 348-2060. Harold J. Sliper, Executive Secretary.

STATE REGISTRATION BOARDS / 295

TENNESSEE: State Board of Architectural and Engineering Examiners, 550 Capitol Hill Building, 301 7th Avenue, North, Nashville, 37219. Telephone: Area code 615, 741-3221. John B. Salsbery, Executive Assistant.

TEXAS: State Board of Registration for Professional Engineers, Room 200, 1400 Congress, Austin, 78701. Telephone: Area code 512, 475-3141. Donald C. Klein, Executive Director. State Board of Registration for Public Surveyors, Room 511, Sam Houston State Office Building, Austin, 78701. Telephone: Area code 512, 475-3445. Betty J. Pope, Executive Secretary.

UTAH: Representative Committee of Professional Engineers and Land Surveyors, 330 East 4th South, Room 210, Salt Lake City, 84111. Telephone: Area code 801, 328-5711. Floy W. McGinn, Director, Department of Registration.

VERMONT: State Board of Registration for Professional Engineers, Norwich University, Northfield, 05663. Telephone: Area code 802, 485-9341. Dr. Walter D. Emerson, P.E., Executive Secretary.

VIRGINIA: State Board for the Examination and Certification of Architects, Professional Engineers and Land Surveyors, PO Box 1-X, Richmond, 23202. Telephone: Area code 703, 770-2161. Ms. Ruth J. Herrick, Secretary.

VIRGIN ISLANDS (US): Board for Architects, Engineers, and Land Surveyors, Dept. of Public Works, PO Box 476, St. Thomas, 00801. John E. Harding, Secretary.

WASHINGTON: State Board of Professional Engineers and Land Surveyors, Division of Professional Licensing, PO Box 649, Olympia, 98504. Telephone: Area code 206, 753-6966. Quentin H. Gateley, P.E., Executive Secretary.

WEST VIRGINIA: State Board of Registration for Professional Engineers, Room 411, 1800 East Washington St., Charleston, 25305. Telephone: Area code 304, 348-3554. William E. Moore II, Secretary.

WISCONSIN: Examining Board of Architects, Professional Engineers, Designers, and Land Surveyors, 201 E. Washington Avenue, Room 252, Madison, 53702. Telephone: Area code 608, 266-1397. Cass F. Hurc, P.E., Administrator Secretary.

WYOMING: State Board of Examining Engineers, State Office Building, Cheyene, 82001. Telephone: Area code 307, nos. 777-7354, 777-7317, or 777-7355. Elva Myers, Secretary-Accountant.

Appendix 4
TEST–TAKING MADE SIMPLE

Having gotten this far, you're almost an expert test-taker because you have now mastered the subject matter of the test. Proper preparation is the real secret. The pointers on the next few pages will take you the rest of the way by giving you the strategy employed on tests by those who are most successful in this not-so-mysterious art.

BEFORE THE TEST

T-DAY MINUS SEVEN

You're going to pass this examination because you have received the best possible preparation for it. But, unlike many others, you're going to give the best possible account of yourself by acquiring the rare skill of effectively using your knowledge to answer the examination questions.

First off, get rid of any negative attitudes toward the test. You have a negative attitude when you view the test as a device to "trip you up" rather than an opportunity to show how effectively you have learned.

APPROACH THE TEST WITH SELF-CONFIDENCE.
Plugging through this book was no mean job, and now that you've done it you're probably better prepared than 90% of the others. Self-confidence is one of the biggest strategic assets you can bring to the testing room.

Nobody likes tests, but some poor souls permit themselves to get upset or angry when they see what they think is an unfair test. The expert doesn't. He keeps calm and moves right ahead, knowing that everyone is taking the same test. Anger, resentment, fear . . . they all slow you down. "Grin and bear it!"

Besides, every test you take, including this one, is a valuable experience which improves your skill. Since you will undoubtedly be taking other tests in the years to come, it may help you to regard this one as training to perfect your skill.

Keep calm; there's no point in panic. If you've done your work there's no need for it; and if you haven't, a cool head is your very first requirement.

Why be the frightened kind of student who enters the examination chamber in a mental coma? A test taken under mental stress does not provide a fair measure of your ability. At the very least, this book has removed for you some of the fear and mystery that surrounds examinations. A certain amount of concern is normal and good, but excessive worry saps your strength and keenness. In other words, be prepared EMOTIONALLY.

Pre-Test Review

If you know any others who are taking this test, you'll probably find it helpful to review the book and your notes with them. The group should be small, certainly not more than four. Team study at this stage should seek to review the material in a different way than you learned it originally; should strive for an exchange of ideas between you and the other members of the group; should be selective in sticking to important ideas; should stress the vague and the unfamiliar rather than that which you all know well; should be businesslike and devoid of any nonsense; should end as soon as you get tired.

One of the *worst* strategies in test taking is to do *all* your preparation the night before the exam. As a reader of this book, you have scheduled and spaced your study properly so as not to suffer from the fatigue and emotional disturbance that comes from cramming the night before.

Cramming is a very good way to *guarantee poor test results*.

However, you would be wise to prepare yourself factually by *reviewing your notes* in the 48 hours preceding the exam. You shouldn't have to spend more than two or three hours in this way. Stick to salient points. The others will fall into place quickly.

Don't confuse cramming with a final, calm review which helps you focus on the significant areas of this book and further strengthens your confidence in your ability to handle the test questions. In other words, prepare yourself FACTUALLY.

Keep Fit

Mind and body work together. Poor physical condition will lower your mental efficiency. In preparing for an examination, observe the common-sense rules of health. Get sufficient sleep and rest, eat proper foods, plan recreation and exercise. In relation to health and examinations, two cautions are in order. Don't miss your meals prior to an examination in order to get extra time for study. Likewise, don't miss your regular sleep by sitting up late to "cram" for the examination. Cramming is an attempt to learn in a very short period of time what should have been learned through regular and consistent study. Not only are these two habits detrimental to health, but seldom do they pay off in terms of effective learning. It is likely that you will be *more confused* than better prepared on the day of the examination if you have broken into your daily routine by missing your meals or sleep.

On the night before the examination go to bed at your regular time and try to get a good night's sleep. Don't go to the movies. Don't date. In other words, prepare yourself PHYSICALLY.

T-HOUR MINUS ONE

After a very light, leisurely meal, get to the examination room ahead of time, perhaps ten minutes early . . . but not so early that you have time to get into an argument with others about what's going to be asked on the exam, etc. The reason for coming early is to help you get accustomed to the room. It will help you to a better start.

Bring all necessary equipment . . .

. . . pen, two sharpened pencils, watch, paper, eraser, ruler, and any other things you're instructed to bring.

Get settled . . .

. . . by finding your seat and staying in it. If no special seats have been assigned, take one in the front to facilitate the seating of others coming in after you.

The test will be given by a test supervisor who reads the directions and otherwise tells you what to do. The people who walk about passing out the test papers and assisting with the examination are test proctors. If you're not able to see or hear properly notify the supervisor or a proctor. If you have any other difficulties during the examination, like a defective test booklet, scoring pencil, answer sheet; or if it's too hot or cold or dark or drafty, let them know. You're entitled to favorable test conditions, and if you don't have them you won't be able to do your best. Don't be a crank, but don't be shy either. An important function of the proctor is to see to it that you have favorable test conditions.

Relax . . .

. . . and don't bring on unnecessary tenseness by worrying about the difficulty of the examination. If necessary wait a minute before beginning to write. If you're still tense, take a couple of deep breaths, look over your test equipment, or do something which will take your mind away from the examination for a moment.

If your collar or shoes are tight, loosen them.

Put away unnecessary materials so that you have a good, clear space on your desk to write freely.

You Must Have
TO GIVE YOUR Best Test PERFORMANCE

(1) A GOOD TEST ENVIRONMENT

(2) A COMPLETE UNDERSTANDING OF DIRECTIONS

(3) A DESIRE TO DO YOUR BEST

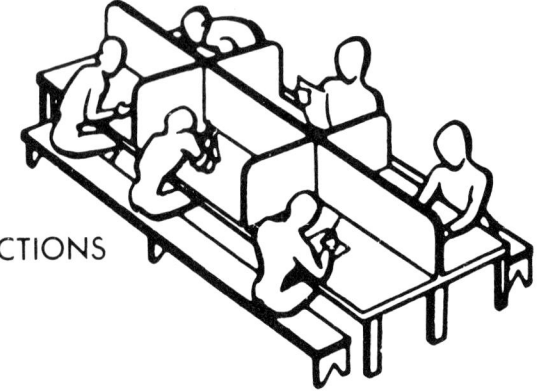

WHEN THEY SAY "GO" — TAKE YOUR TIME!

Listen very carefully to the test supervisor. If you fail to hear something important that he says, you may not be able to read it in the written directions and may suffer accordingly.

If you don't understand the directions you have heard or read, raise your hand and inform the proctor. Read carefully the directions for *each* part of the test before beginning to work on that part. If you skip over such directions too hastily, you may miss a main idea and thus lose credit for an entire section.

Get an Overview of the Examination

After reading the directions carefully, look over the entire examination to get an over-view of the nature and scope of the test. The purpose of this over-view is to give you some idea of the nature, scope, and difficulty of the examination.

It has another advantage. An item might be so phrased that it sets in motion a chain of thought that might be helpful in answering other items on the examination.

Still another benefit to be derived from reading all the items before you answer any is that the few minutes involved in reading the items gives you an opportunity to relax before beginning the examination. This will make for better concentration. As you read over these items the first time, check those whose answers immediately come to you. These will be the ones you will answer first. Read each item carefully before answering. It is a good practice to read each item at least twice to be sure that you understand it.

Plan Ahead

In other words, you should know precisely where you are going before you start. You should know:
1. whether you have to answer all the questions or whether you can choose those that are easiest for you;
2. whether all the questions are easy; (there may be a pattern of difficult, easy, etc.)
3. The length of the test; the number of questions;
4. The kind of scoring method used;
5. Which questions, if any, carry extra weight;
6. What types of questions are on the test;
7. What directions apply to each part of the test;
8. Whether you must answer the questions consecutively.

Budget Your Time Strategically!

Quickly figure out how much of the allotted time you can give to each section and still finish ahead of time. Don't forget to figure on the time you're investing in the overview. Then alter your schedule so that you can spend more time on those parts that count most. Then, if you can, plan to spend less time on the easier questions, so that you can devote the time saved to the harder questions. Figuring roughly, you should finish half the questions when half the allotted time has gone by. If there are 100 questions and you have three hours, you should have finished 50 questions after one and one half hours. So bring along a watch whether the instructions call for one or not. Jot down your "exam budget" and stick to it INTELLIGENTLY.

EXAMINATION STRATEGY

Probably the most important single strategy you can learn is to do the easy questions first. The very hard questions should be read and temporarily postponed. Identify them with a dot and return to them later.

This strategy has several advantages for you:
1. You're sure to get credit for all the questions you're sure of. If time runs out, you'll have all the sure shots, losing out only on those which you might have missed anyway.

2. By reading and laying away the tough ones you give your subconscious a chance to work on them. You may be pleasantly surprised to find the answers to the puzzlers popping up for you as you deal with related questions.

3. You won't risk getting caught by the time limit just as you reach a question you know really well.

A Tested Tactic

It's inadvisable on some examinations to answer each question in the order presented. The reason for this is that some examiners design tests so as to extract as much mental energy from you as possible. They put the most difficult questions at the beginning, the easier questions last. Or they may vary difficult with easy questions in a fairly regular pattern right through the test. Your survey of the test should reveal the pattern and your strategy for dealing with it.

If difficult questions appear at the beginning, answer them until you feel yourself slowing down or getting tired. Then switch to an easier part of the examination. You will return to the difficult portion after you have rebuilt your confidence by answering a batch of easy questions. Knowing that you have a certain number of points "under your belt" will help you when you return to the more difficult questions. You'll answer them with a much clearer mind; and you'll be refreshed by the change of pace.

Time

Use your time wisely. It's an important element in your test and you must use every minute effectively, working as rapidly as you can without sacrificing accuracy. Your exam survey and budget will guide you in dispensing your time. Wherever you can, pick up seconds on the easy ones. Devote your savings to the hard ones. If possible, pick up time on the lower value questions and devote it to those which give you the most points.

Relax Occasionally and Avoid Fatigue

If the exam is long (two or more hours) give yourself short rest periods as you feel you need them. If you're not permitted to leave the room, relax in your seat, look up from your paper, rest your eyes, stretch your legs, shift your body. Break physical and mental tension. Take several deep breaths and get back to the job, refreshed. If you don't do this you run the risk of getting nervous and tightening up. Your thinking may be hampered and you may make a few unnecessary mistakes.

Do not become worried or discouraged if the examination seems difficult to you. The questions in the various fields are purposely made difficult and searching so that the examination will discriminate effectively even among superior students. No one is expected to get a perfect or near-perfect score.

Remember that if the examination seems difficult to you, it may be even more difficult for your neighbor.

Think!

This is not a joke because you're not an IBM machine. Nobody is able to write all the time and also to read and think through each question. You must plan each answer. Don't give hurried answers in an atmosphere of panic. Even though you see a lot of questions, remember that they are objective and not very time-consuming. Don't rush headlong through questions that must be thought through.

Edit, Check, Proofread . . .

. . . after completing all the questions. Invariably, you will find some foolish errors which you needn't have made, and which you can easily correct. Don't just sit back or leave the room ahead of time. Read over your answers and make sure you wrote exactly what you meant to write. And that you wrote the answers in the right place. You might even find that you have omitted some answers inadvertently. You have budgeted time for this job of proofreading. PROOFREAD and pick up points.

One caution, though. Don't count on making major changes. And don't go in for wholesale changing of answers. To arrive at your answers in the first place you have read carefully and thought correctly. Second-guessing at this stage is more likely to result in wrong answers. So don't make changes unless you are quite certain you were wrong in the first place.

FOLLOW DIRECTIONS CAREFULLY

In answering questions on the objective or short-form examination, it is most important to follow all instructions carefully. Unless you have marked the answers properly, you will not receive credit for them. In addition, even in the same examination, the instructions will not be consistent. In one section you may be urged to guess if you are not certain; in another you may be cautioned against guessing. Some questions will call for the best choice among four or five alternatives; others may ask you to select the one incorrect or the least probable answer.

On some tests you will be provided with worked out fore-exercises, complete with correct answers. However, avoid the temptation to skip the directions and begin working just from reading the model questions and answers. Even though you may be familiar with that particular type of question, the directions may be different from those which you had followed previously. If the type of question should be new to you, work through the model until you understand it perfectly. This may save you time, and earn you a higher rating on the examination.

If the directions for the examination are written, read them carefully, at least twice. If the directions are given orally, listen attentively and then follow them precisely. For example, if you are directed to use plus (+) and minus (−) to mark true—false items, then don't use "T" and "F". If you are instructed to "blacken" a space on machine-scored tests, do not use a check (✓) or an "X". Make all symbols legible, and be sure that they have been placed in the proper answer space. It is easy, for example, to place the answer for item 5 in the space reserved for item 6. If this is done, then all of your following answers may be wrong. It is also very important that you understand the method they will use in scoring the examination. Sometimes they tell you in the directions. The method of scoring may affect the amount of time you spend on an item, especially if some items count more than others. Likewise, the directions may indicate whether or not you should guess in case you are not sure of the answer. Some methods of scoring penalize you for guessing.

Cue Words. Pay special attention to qualifying words or phrases in the directions. Such words as *one, best reason, surest, means most nearly the same as, preferable, least correct,* etc., all indicate that *one* response is called for, and that you must select the response which best fits the qualifications in the question.

Time. Sometimes a time limit is set for each section of the examination. If that is the case, follow the time instructions carefully. Your *exam budget* and your watch can help you here. Even if you haven't finished a section when the time limit is up, pass on to the next section. The examination has been planned according to the time schedule. If the examination paper bears the instruction "Do not turn over page until signal is given," or "Do not start until signal is given," follow the instruction. Otherwise, you may be disqualified.

Pay Close Attention. Be sure you understand what you're doing at all times. Especially in dealing with true-false or multiple-choice questions it's vital that you understand the meaning of every question. It is normal to be working under stress when taking an examination, and it is easy to skip a word or jump to a false conclusion, which may cost you points on the examination. In many multiple-choice and matching questions, the examiners deliberately insert plausible-appearing false answers in order to catch the candidate who is not alert.

Answer clearly. If the examiner who marks your paper cannot understand what you mean, you will not receive credit for your correct answer. On a True-False examination you will not receive any credit for a question which is marked both true and false. If you are asked to underline, be certain that your lines are under and not through the words and that they do not extend beyond them. When using the separate answer sheet it is important *when you decide to change an answer,* you erase the first answer completely. If you leave any graphite from the pencil on the wrong space it will cause the scoring machine to cancel the right answer for that question.

Watch Your "Weights." If the examination is "weighted" it means that some parts of the examination are considered more important than others and rated more highly. For instance, you may find that the instructions will indicate "Part I, Weight 50; Part II, Weight 25, Part III, Weight 25." In such a case, you would devote half of your time to the first part, and divide the second half of your time among Parts II and III.

A Funny Thing . . .

. . . happened to you on your way to the bottom of the totem pole. You *thought* the right answer but you marked the *wrong* one.

1. You *mixed answer symbols!* You decided (rightly) that Baltimore (Choice D) was correct. Then you marked *B* (for Baltimore) instead of *D*.

2. You *misread* a simple instruction! Asked to give the *latest* word in a scrambled sentence, you correctly arranged the sentence, and then marked the letter corresponding to the *earliest* word in that miserable sentence.

3. You *inverted digits!* Instead of the correct number, 96, you wrote (or read) 69.

Funny? Tragic! Stay away from accidents.

Record your answers on the answer sheet one by one as you answer the questions. Care should be taken that these answers are recorded next to the appropriate numbers on your answer sheet. It is poor practice to write your answers first on the test booklet and then to transfer them all at one time to the answer sheet. This procedure causes many errors. And then, how would you feel if you ran out of time before you had a chance to transfer all the answers.

When and How To Guess

Read the directions carefully to determine the scoring method that will be used. In some tests, the directions will indicate that guessing is advisable if you do not know the answer to a question. In such tests, only the right answers are counted in determining your score. If such is the case, don't omit any items. If you do not know the answer, or if you are not sure of your answer, then *guess*.

On the other hand, if the directions state that a scoring formula *will* be used in determining your score or that you are *not to guess*, then *omit* the question if you do not know the answer, or if you are not sure of the answer. When the scoring formula is used, a percentage of the *wrong* answers will be subtracted from the number of *right* answers as a correction for haphazard guessing. It is improbable, therefore, that mere guessing will improve your score significantly. *It may even lower your score.* Another disadvantage in guessing under such circumstances is that it consumes valuable time that you might profitably use in answering the questions you know.

If, however, you are uncertain of the correct answer but have *some* knowledge of the question and are able to eliminate one or more of the answer choices as wrong, your chance of getting the right answer is improved, and it will be to your advantage to *answer* such a question rather than *omit* it.

BEAT THE ANSWER SHEET

Even though you've had plenty of practice with the answer sheet used on machine-scored examinations, we must give you a few more, last-minute pointers.

The present popularity of tests requires the use of electrical test scoring machines. With these machines, scoring which would require the labor of several men for hours can be handled by one man in a fraction of the time.

The scoring machine is an amazingly intricate and helpful device, but the machine is not human. The machine cannot, for example, tell the difference between an intended answer and a stray pencil mark, and will count both indiscriminately. The machine cannot count a pencil mark, if the pencil mark is not brought in contact with the electrodes. For these reasons, specially printed answer sheets with response spaces properly located and properly filled in must be employed. Since not all pencil leads contain the necessary ingredients, a special pencil must be used and a heavy solid mark must be made to indicate answers.

(a) Each pencil mark must be heavy and black. Light marks should be retraced with the special pencil.

(b) Each mark must be in the space between the pair of dotted lines and entirely fill this space.

(c) All stray pencil marks on the paper, clearly not intended as answers, must be completely erased.

(d) Each question must have only one answer indicated. If multiple answers occur, all extraneous marks should be thoroughly erased. Otherwise, the machine will give you *no* credit for your correct answer.

Be sure to use the special electrographic pencil!

HERE'S HOW TO MARK YOUR ANSWERS ON MACHINE-SCORED ANSWER SHEETS:

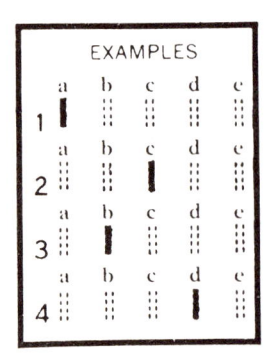

Make only ONE mark for each answer. Additional and stray marks may be counted as mistakes. In making corrections, erase errors COMPLETELY. Make glossy black marks.

Your answer sheet is the only one that reaches the office where papers are scored. For this reason it is important that the blanks at the top be filled in completely and correctly. The proctors will check this, but just in case they slip up, make certain yourself that your paper is complete.

Many exams caution competitors against making any marks on the test booklet itself. Obey that caution even though it goes against your grain to work neatly. If you work neatly and obediently with the test booklet you'll probably do the same with the answer sheet. And that pays off in high scores.

THE GIST OF TEST STRATEGY

- APPROACH THE TEST CONFIDENTLY. TAKE IT CALMLY.
- REMEMBER TO REVIEW, THE WEEK BEFORE THE TEST.
- DON'T "CRAM." BE CAREFUL OF YOUR DIET AND SLEEP. ESPECIALLY AS THE TEST DRAWS NIGH.
- ARRIVE ON TIME... AND READY.
- CHOOSE A GOOD SEAT. GET COMFORTABLE AND RELAX.
- BRING THE COMPLETE KIT OF "TOOLS" YOU'LL NEED.
- LISTEN CAREFULLY TO ALL DIRECTIONS.
- APPORTION YOUR TIME INTELLIGENTLY WITH AN "EXAM BUDGET."
- READ ALL DIRECTIONS CAREFULLY. TWICE IF NECESSARY. PAY PARTICULAR ATTENTION TO THE SCORING PLAN.
- LOOK OVER THE WHOLE TEST BEFORE ANSWERING ANY QUESTIONS.
- START RIGHT IN, IF POSSIBLE. STAY WITH IT. USE EVERY SECOND EFFECTIVELY.
- DO THE EASY QUESTIONS FIRST; POSTPONE HARDER QUESTIONS UNTIL LATER.
- DETERMINE THE PATTERN OF THE TEST QUESTIONS. IF IT'S HARD-EASY ETC., ANSWER ACCORDINGLY.
- READ EACH QUESTION CAREFULLY. MAKE SURE YOU UNDERSTAND EACH ONE BEFORE YOU ANSWER. RE-READ, IF NECESSARY.
- THINK! AVOID HURRIED ANSWERS. GUESS INTELLIGENTLY.
- WATCH YOUR WATCH AND "EXAM BUDGET," BUT DO A LITTLE BALANCING OF THE TIME YOU DEVOTE TO EACH QUESTION.
- GET ALL THE HELP YOU CAN FROM "CUE" WORDS.
- REPHRASE DIFFICULT QUESTIONS FOR YOURSELF. WATCH OUT FOR "SPOILERS."
- REFRESH YOURSELF WITH A FEW, WELL-CHOSEN REST PAUSES DURING THE TEST.
- USE CONTROLLED ASSOCIATION TO SEE THE RELATION OF ONE QUESTION TO ANOTHER AND WITH AS MANY IMPORTANT IDEAS AS YOU CAN DEVELOP.
- NOW THAT YOU'RE A "COOL" TEST-TAKER, STAY CALM AND CONFIDENT THROUGHOUT THE TEST. DON'T LET ANYTHING THROW YOU.
- EDIT, CHECK, PROOFREAD YOUR ANSWERS. BE A "BITTER ENDER." STAY WORKING UNTIL THEY MAKE YOU GO.

HOW TO BE A MASTER TEST TAKER

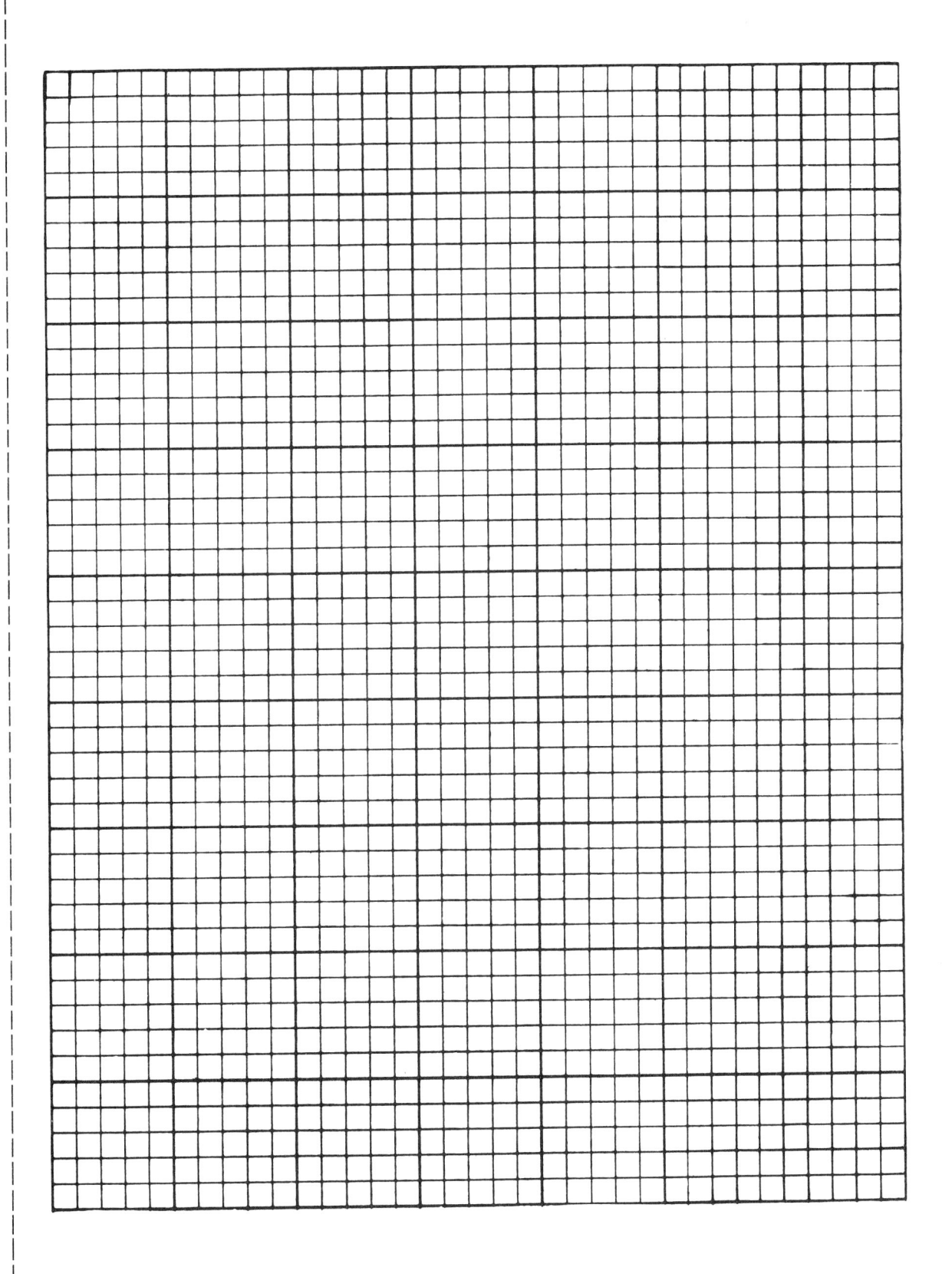

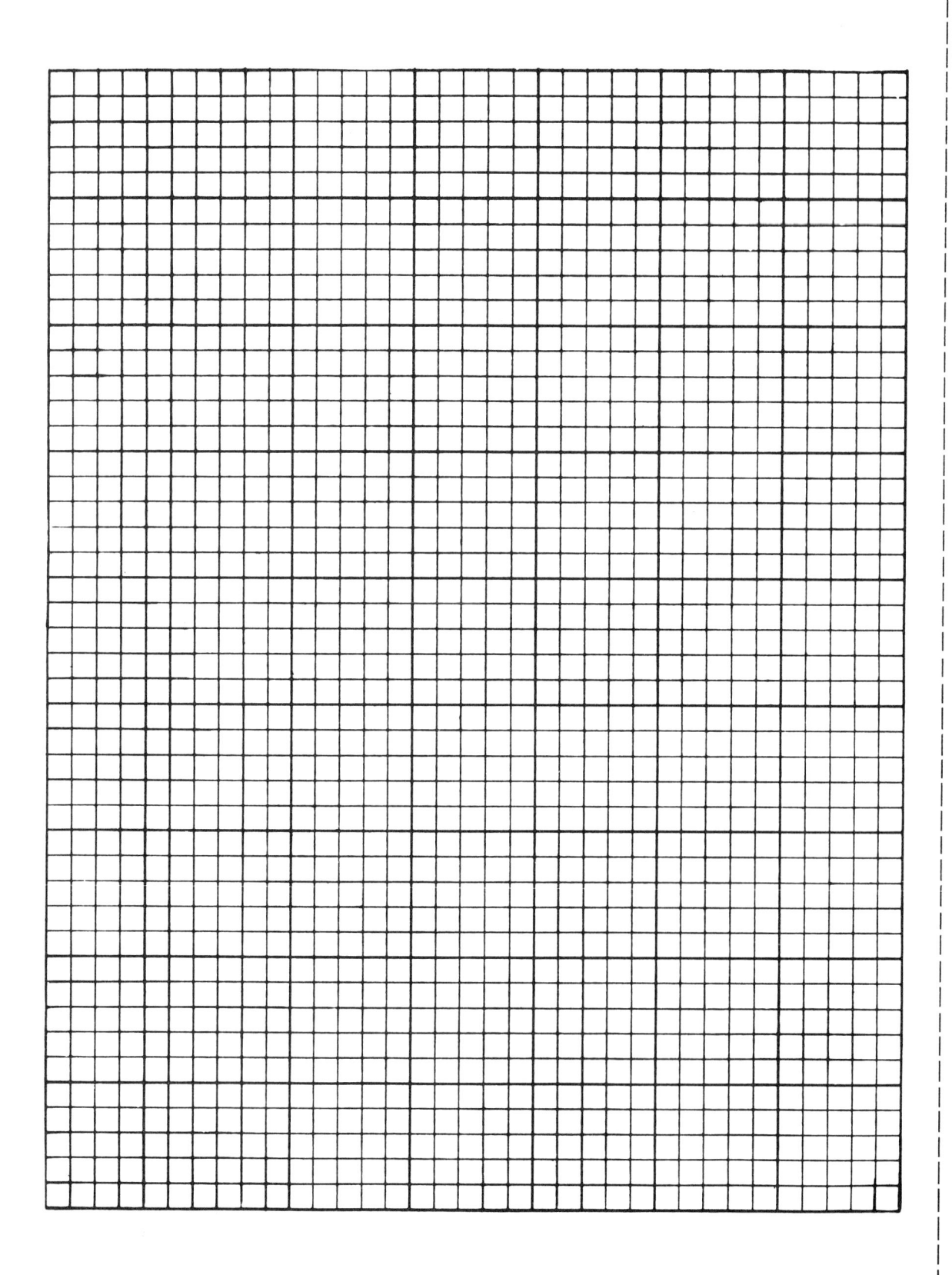

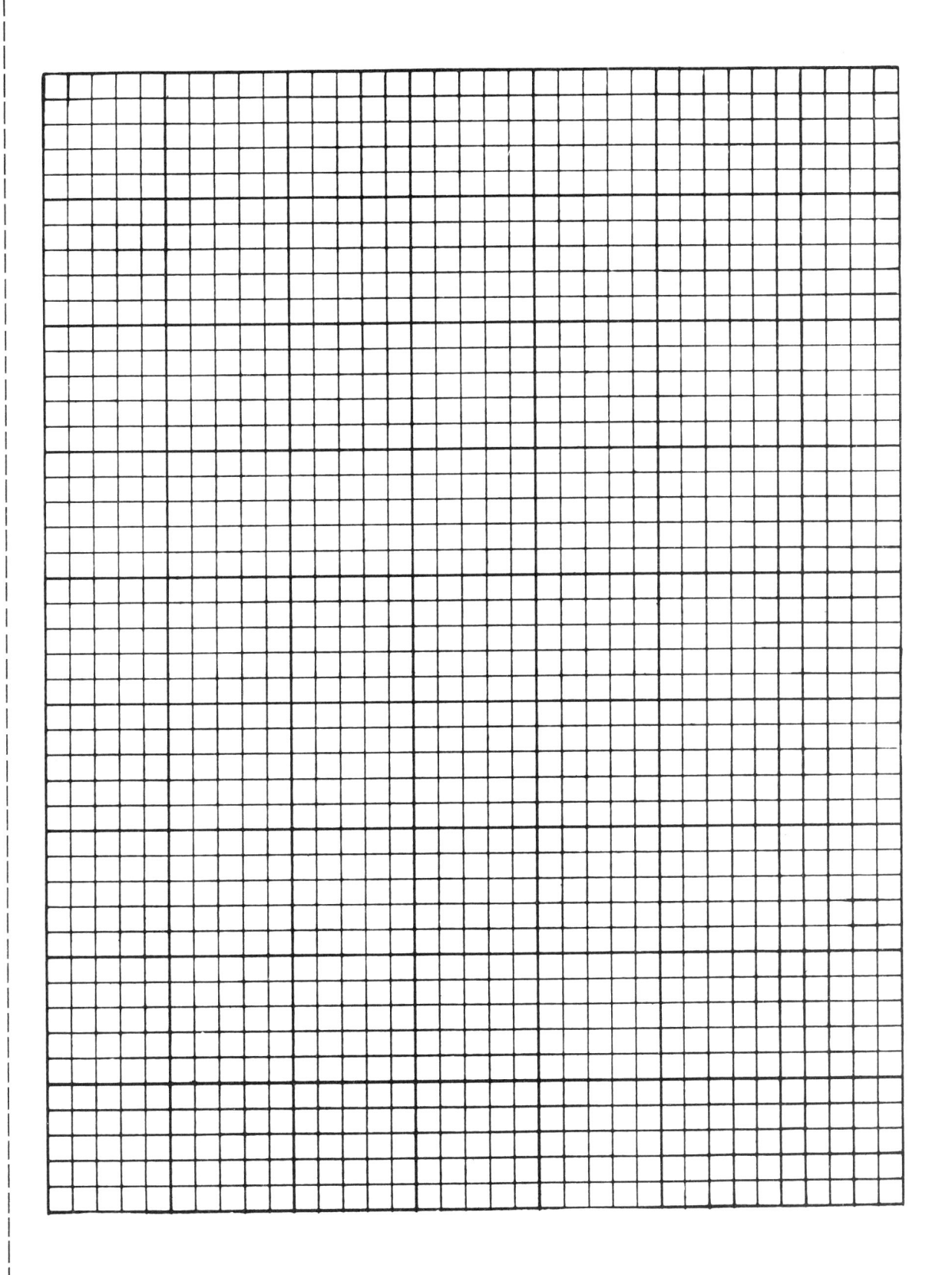

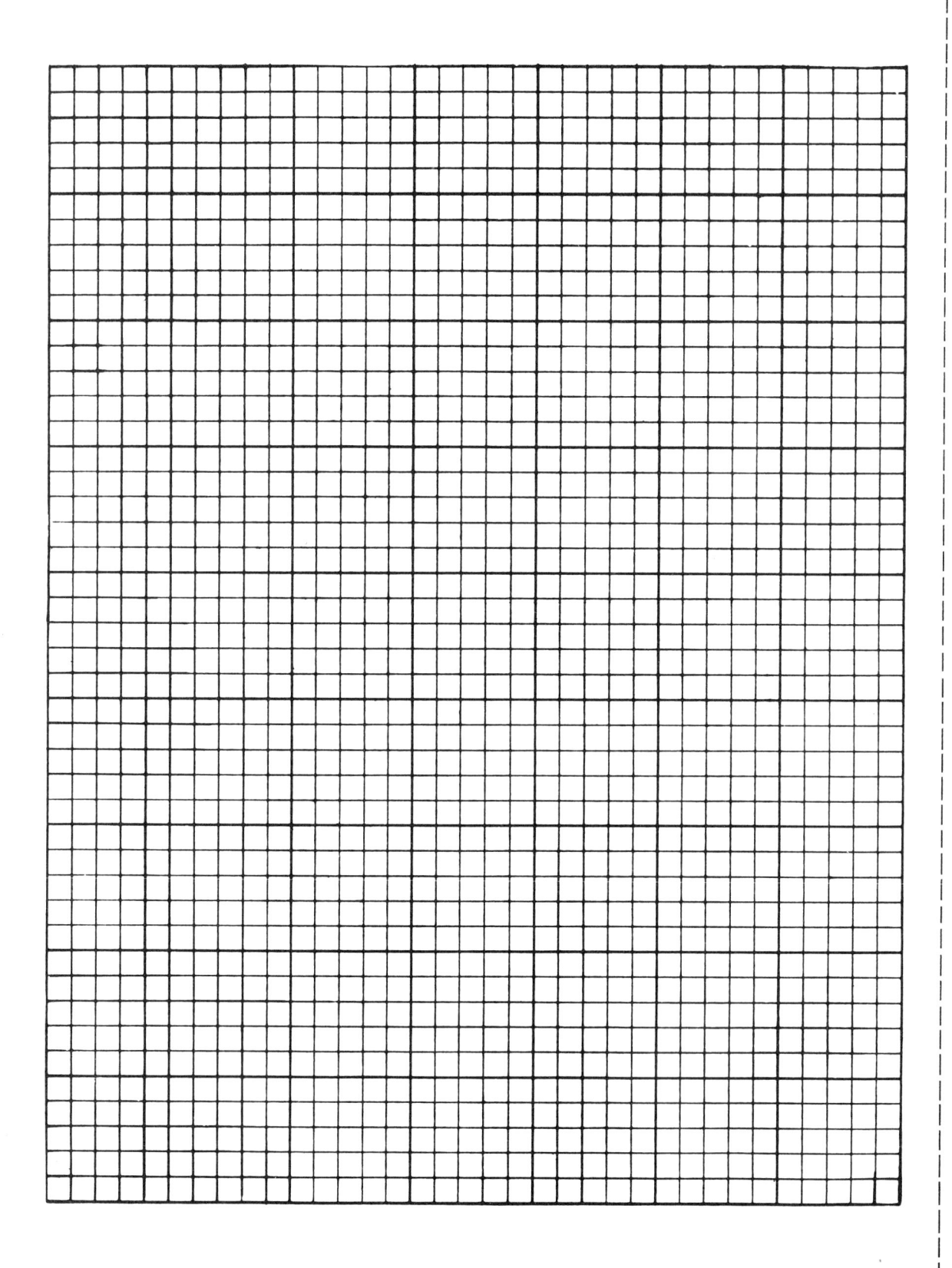

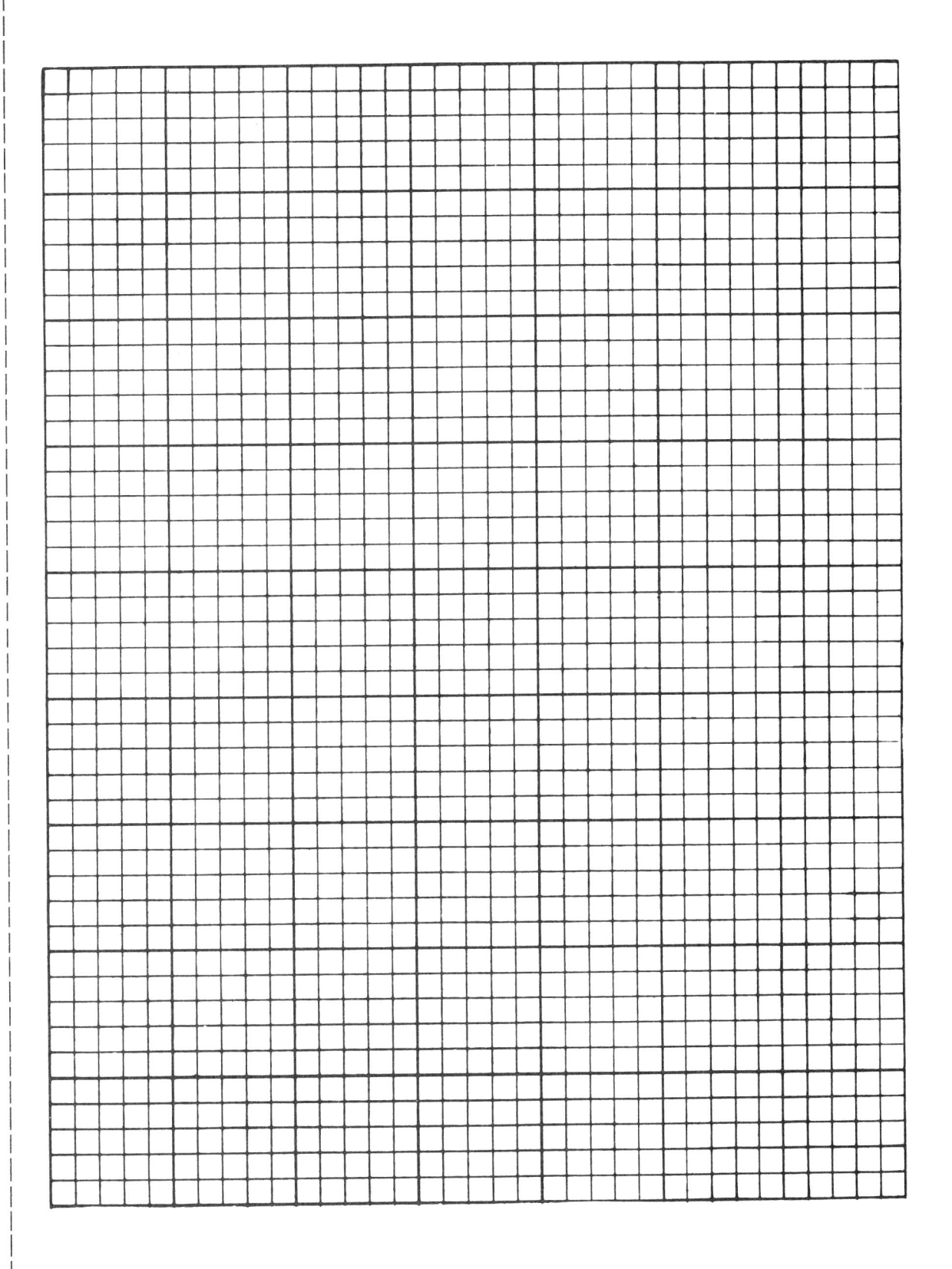

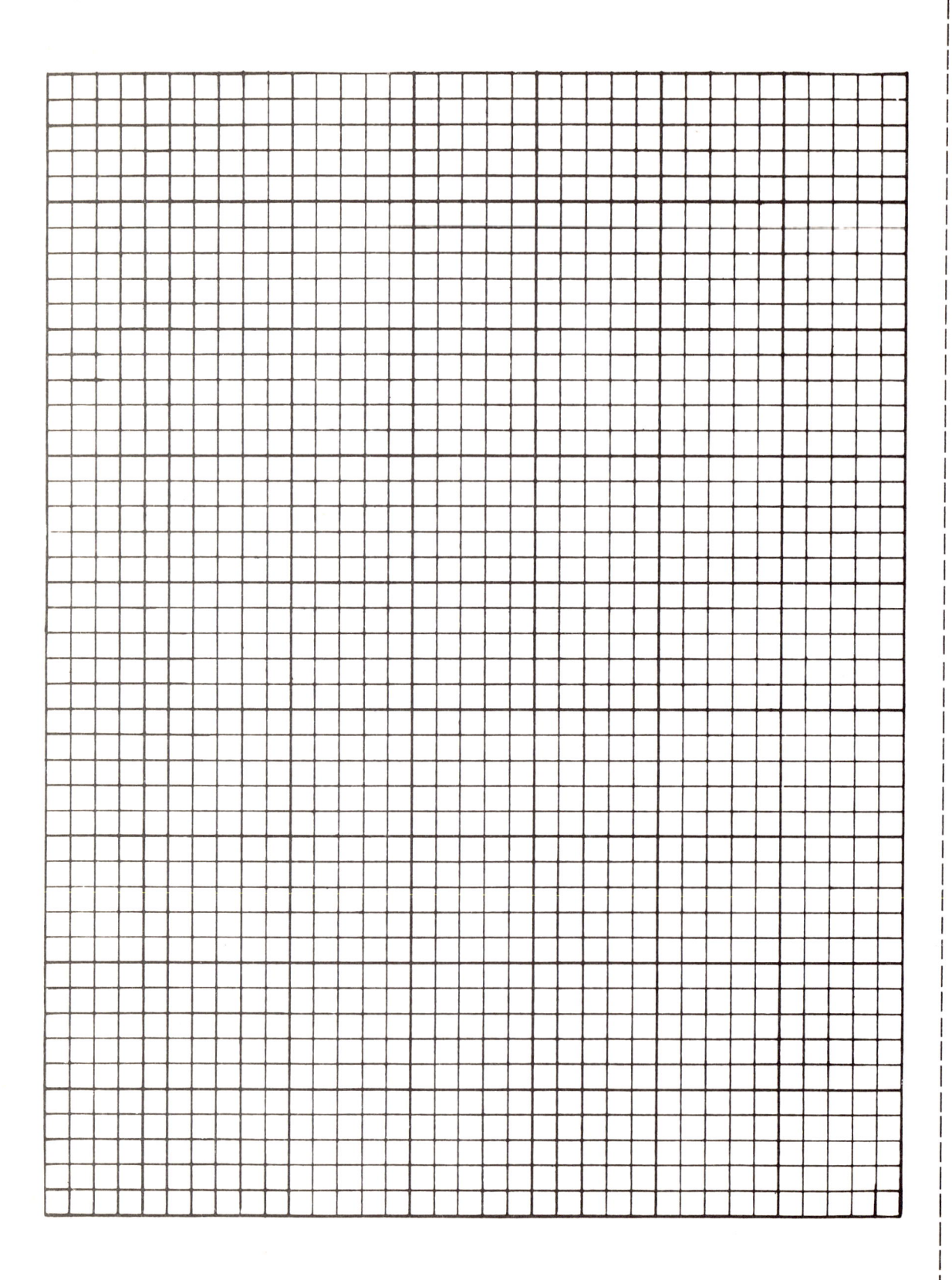

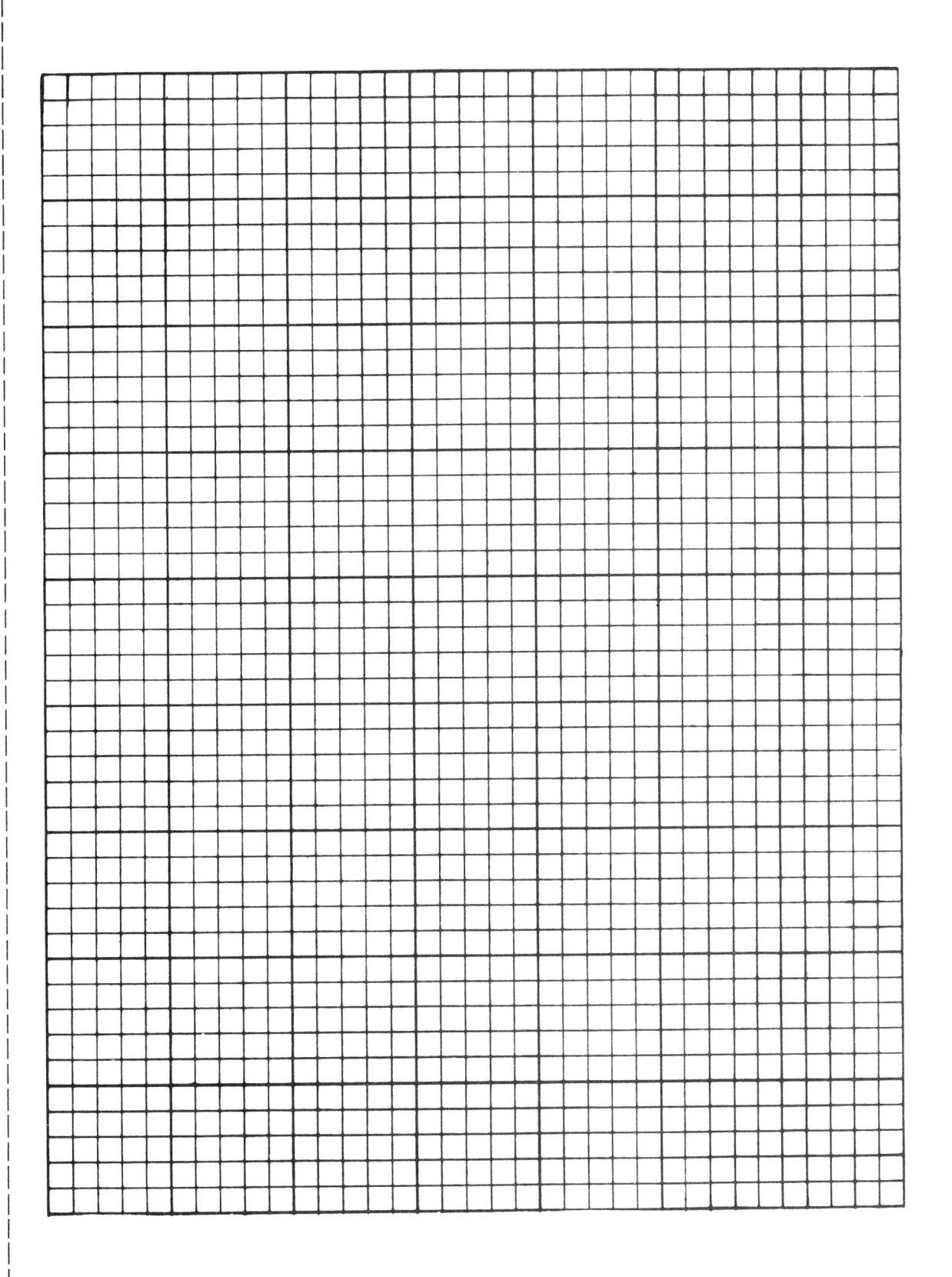

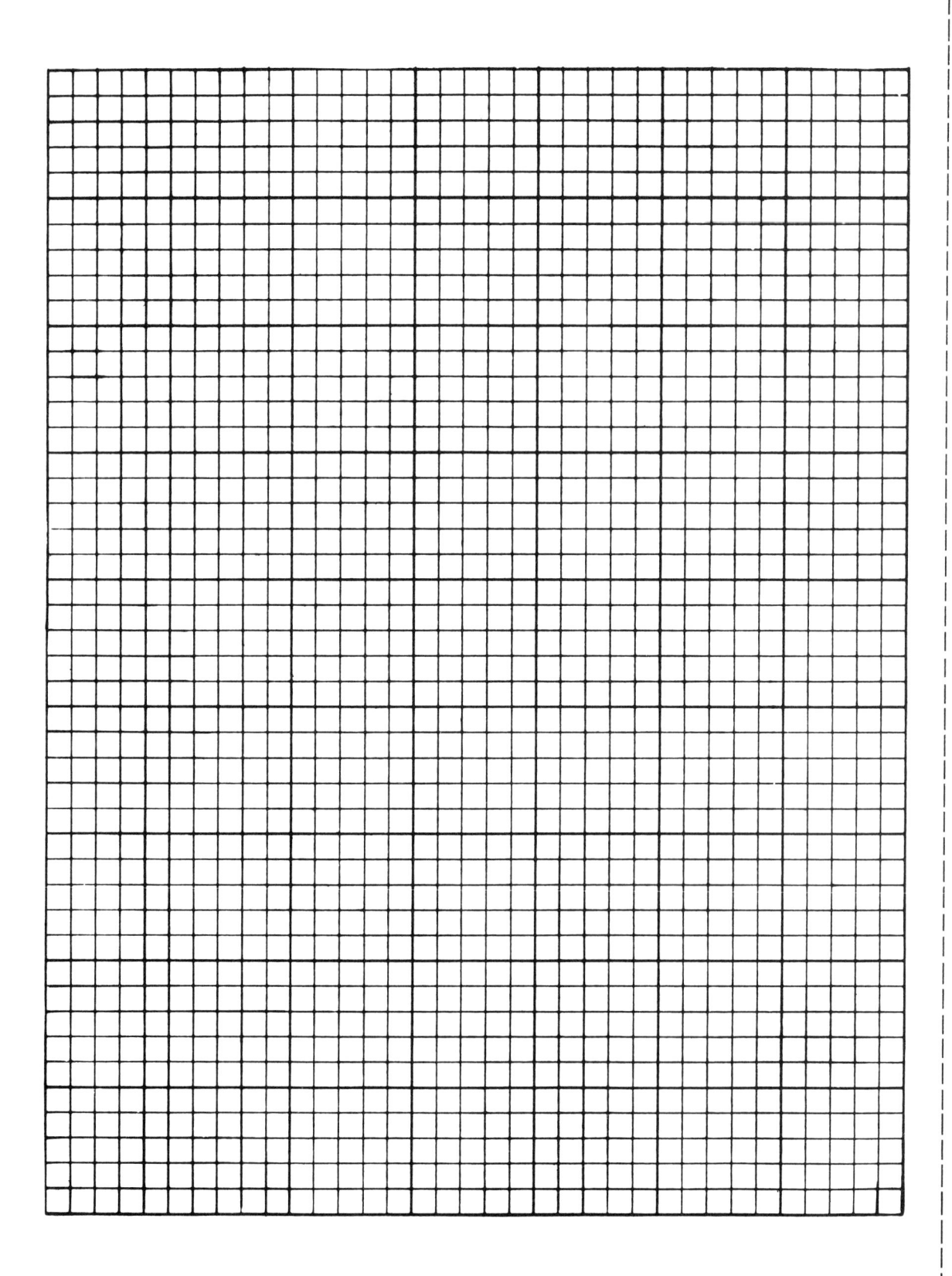

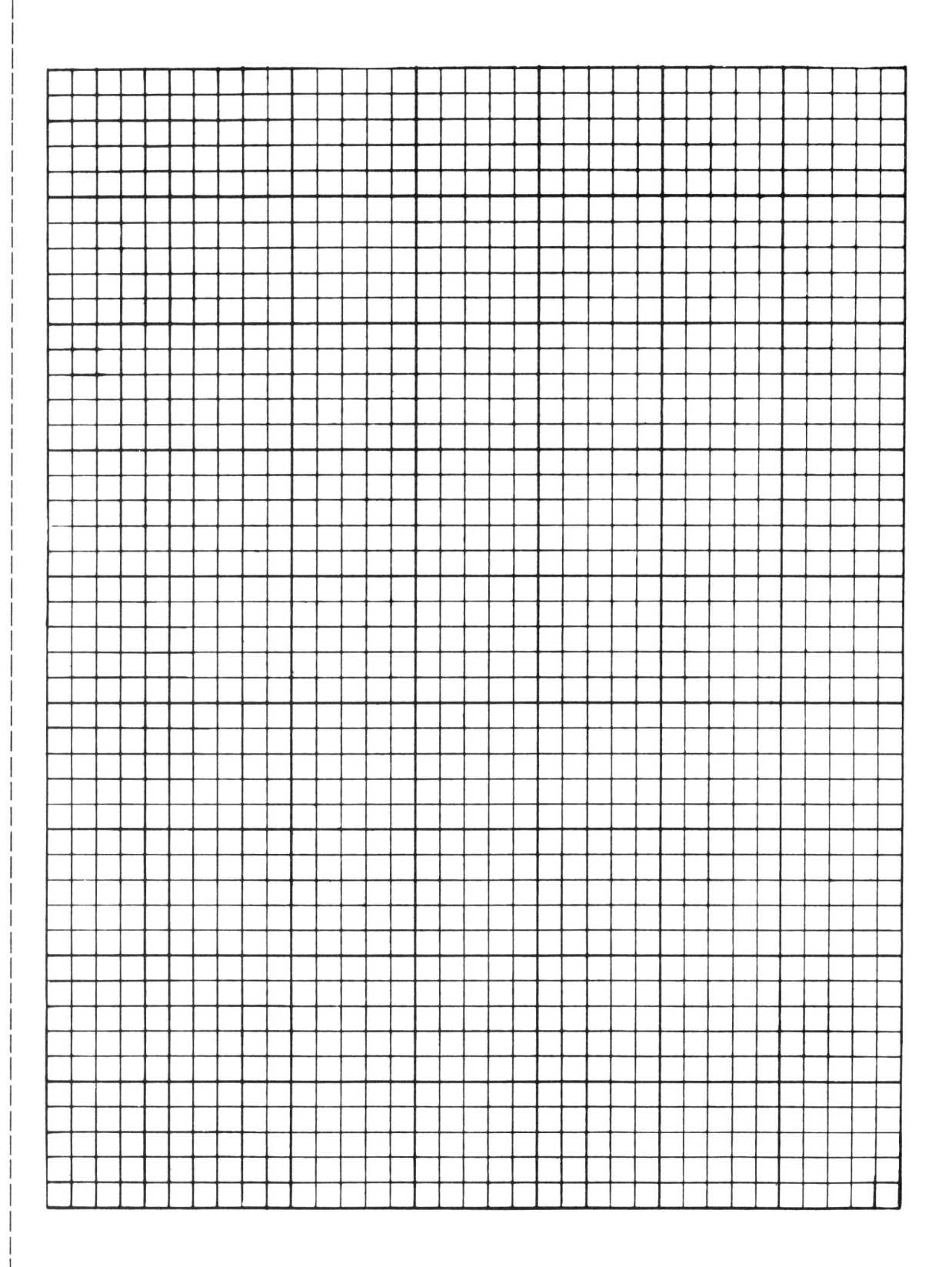

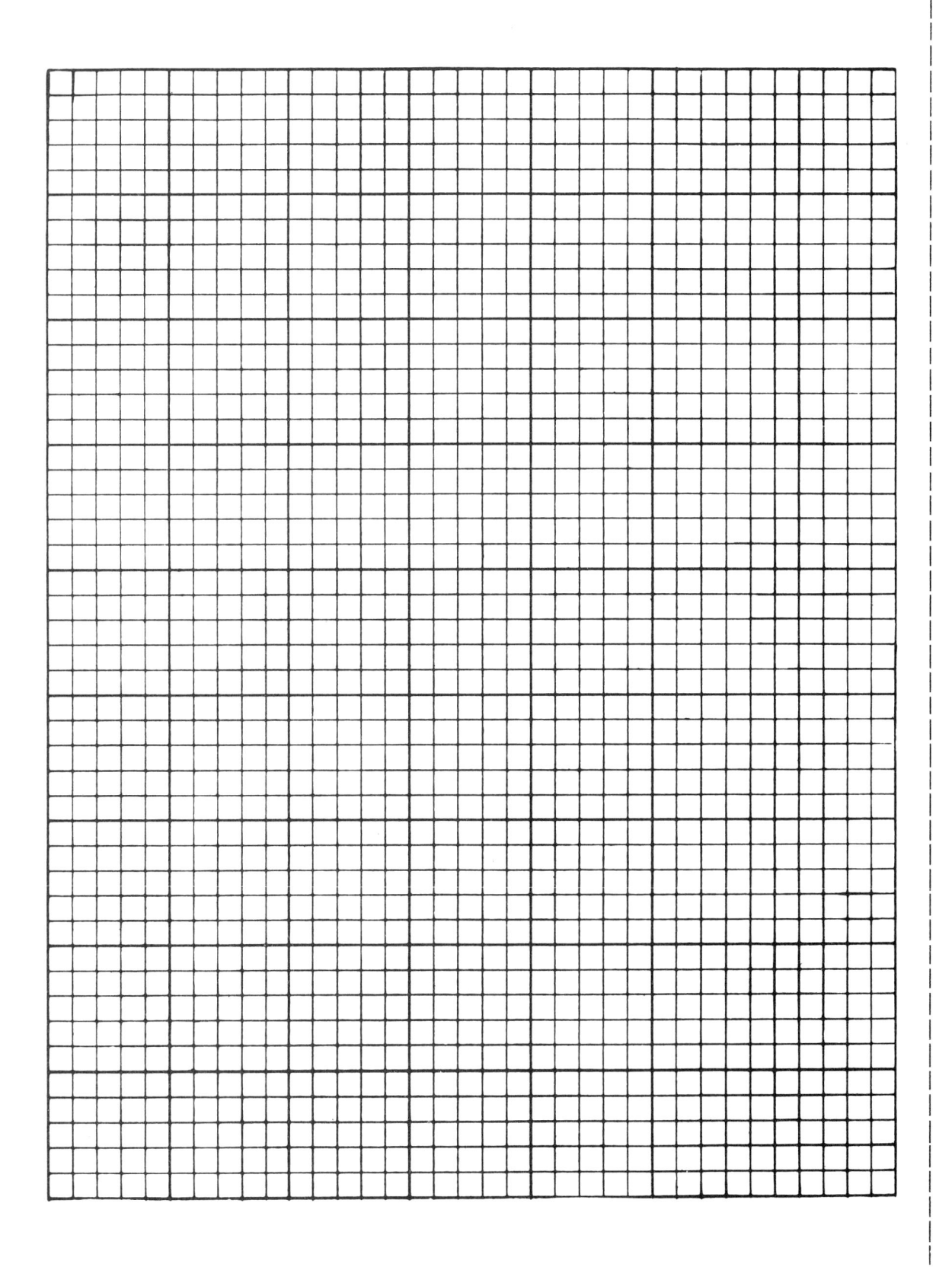

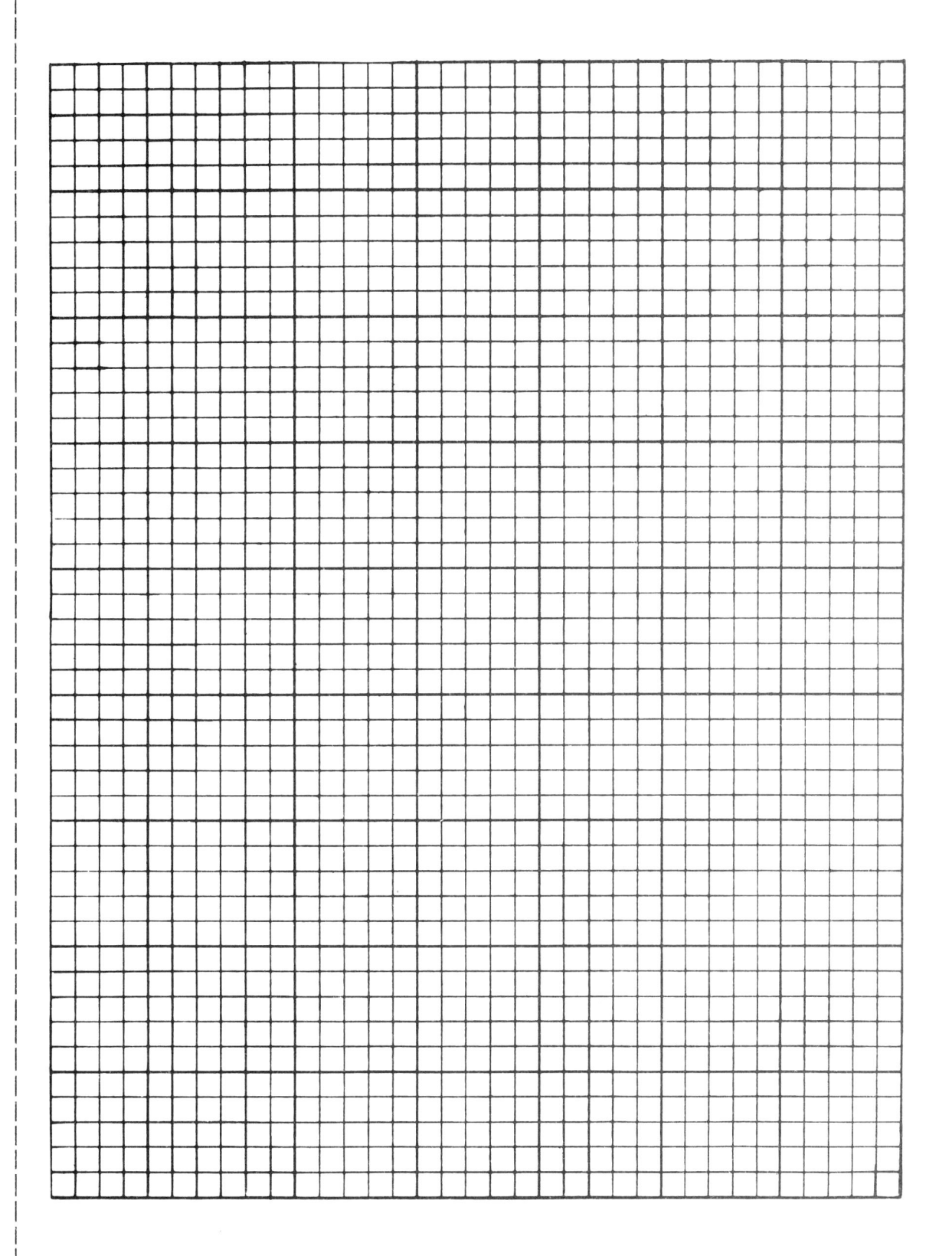

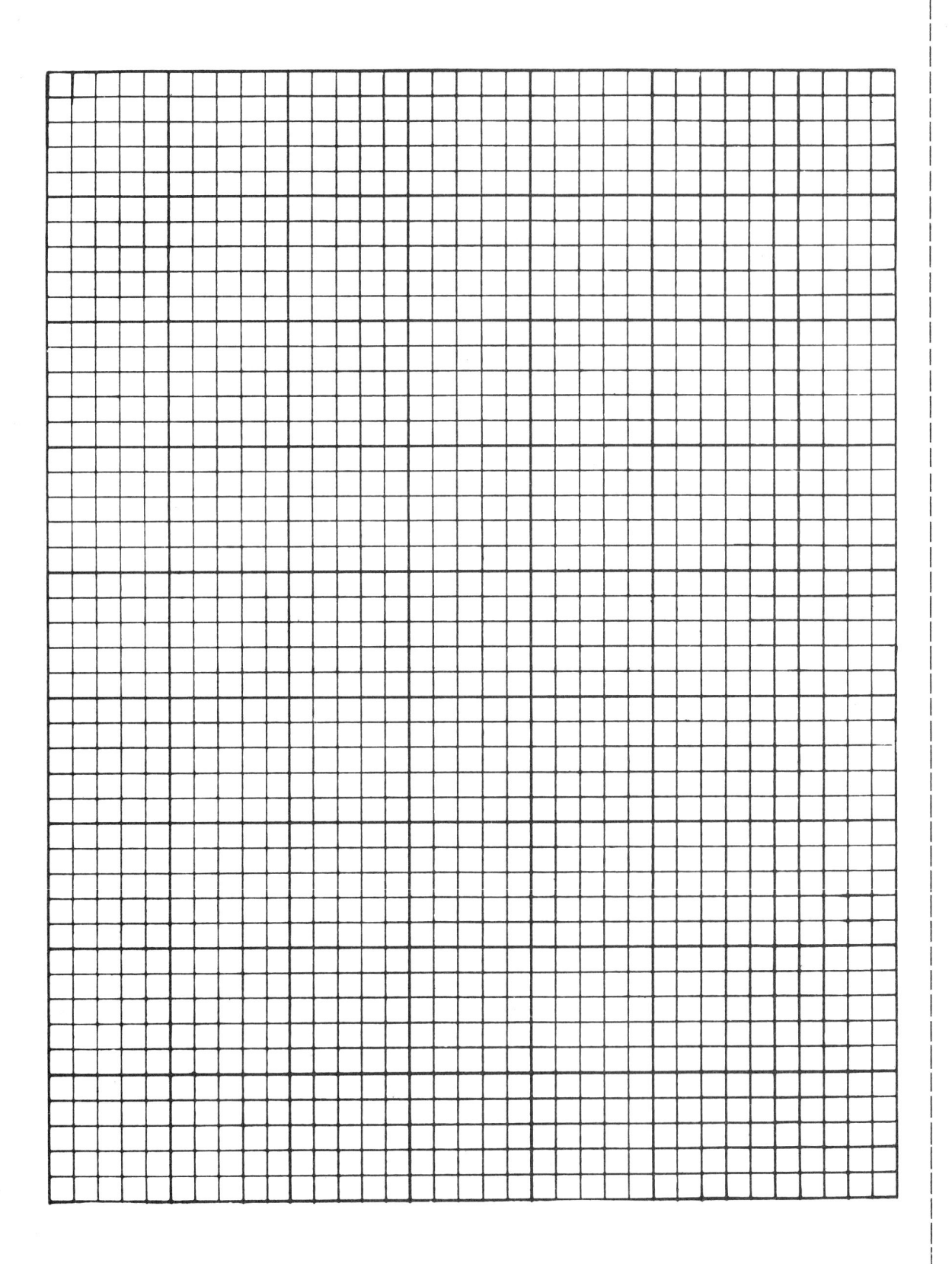

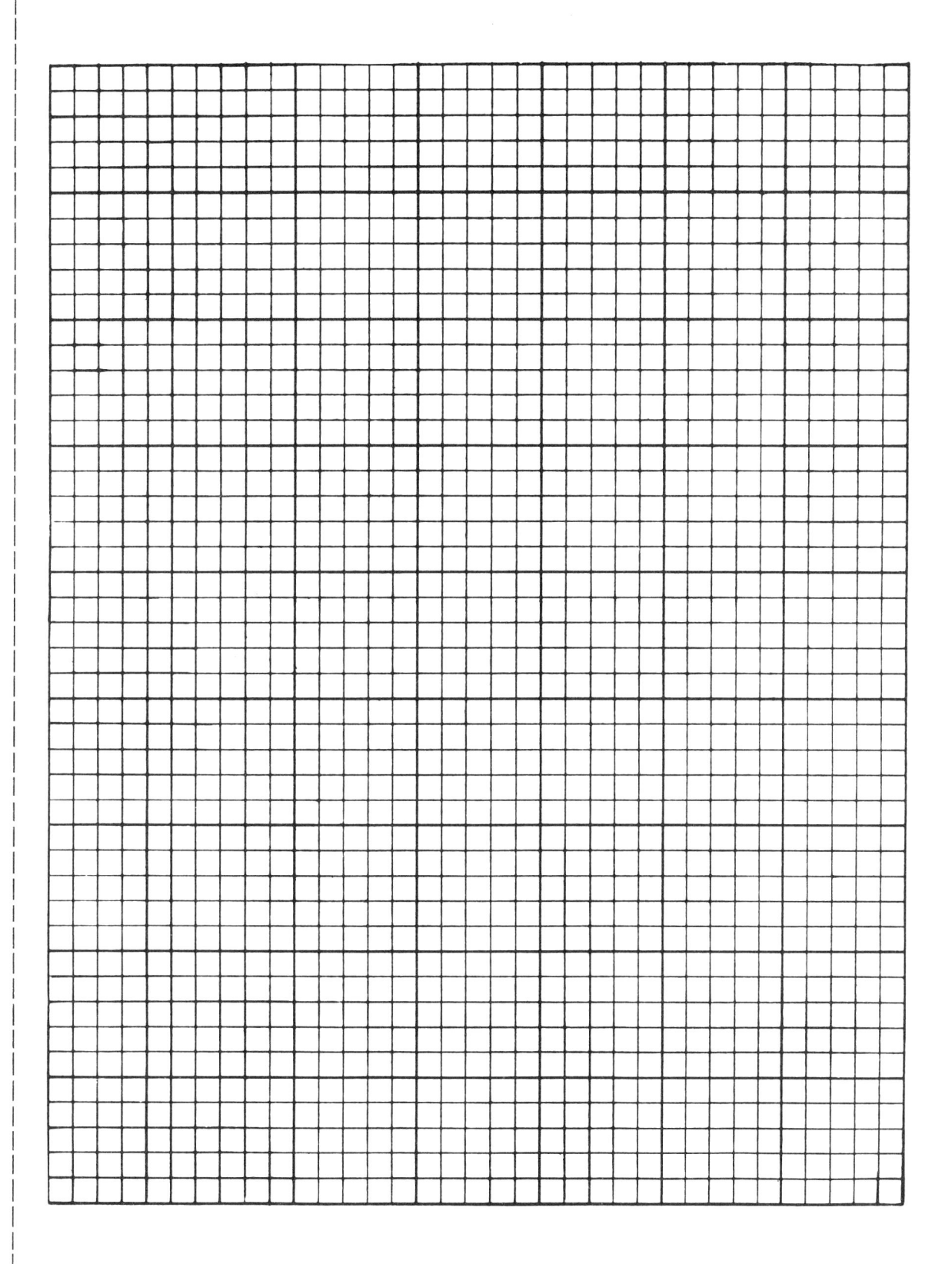

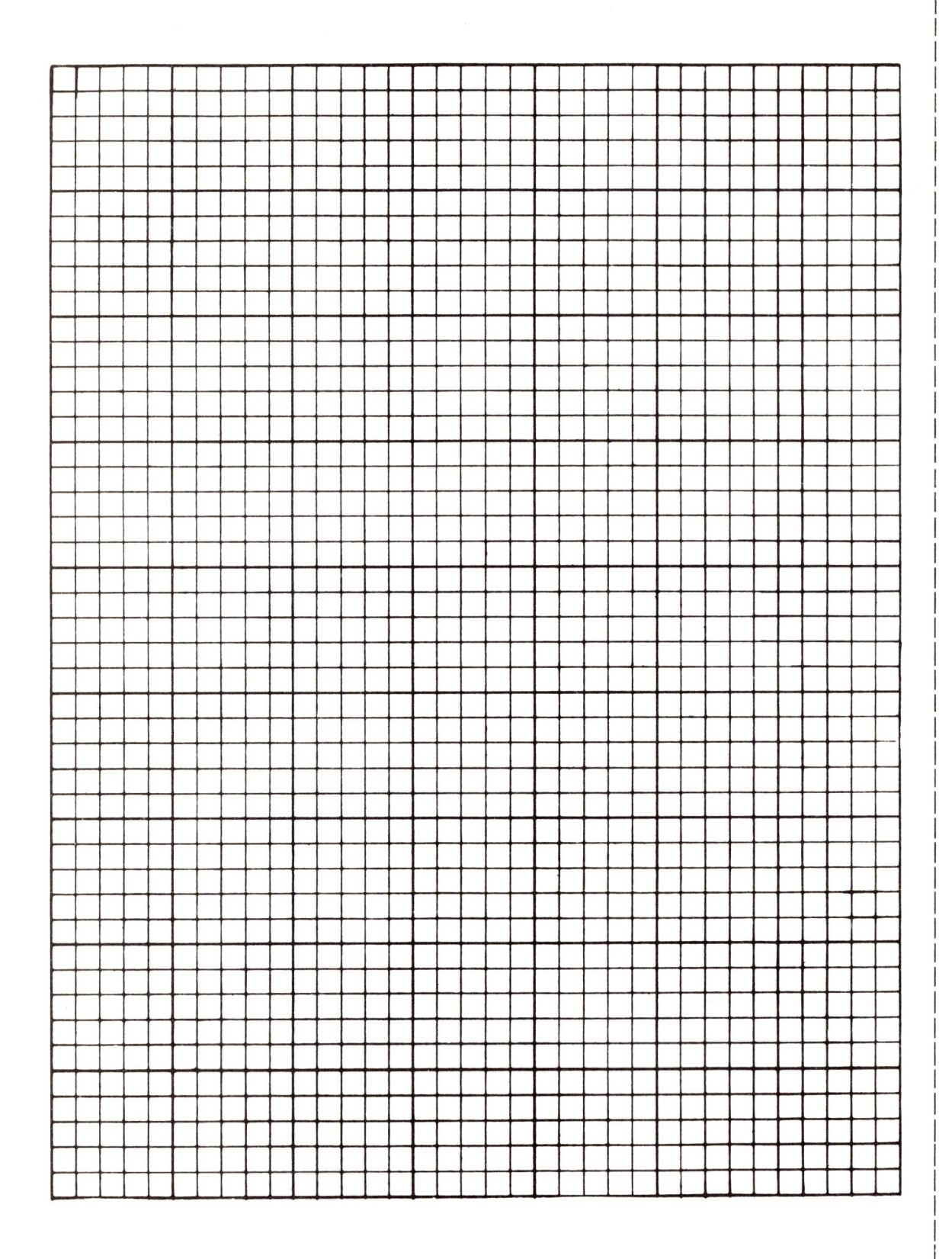

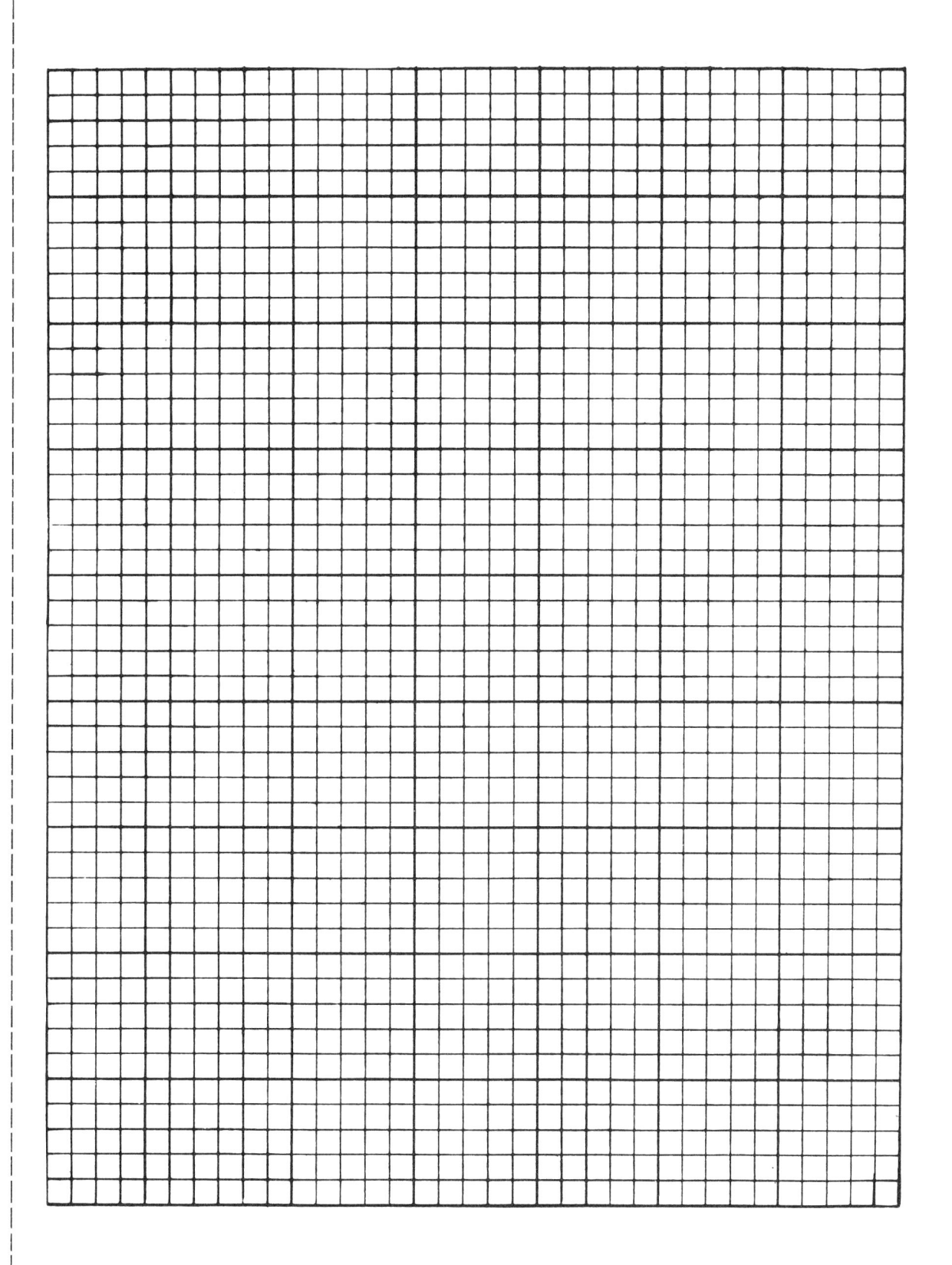

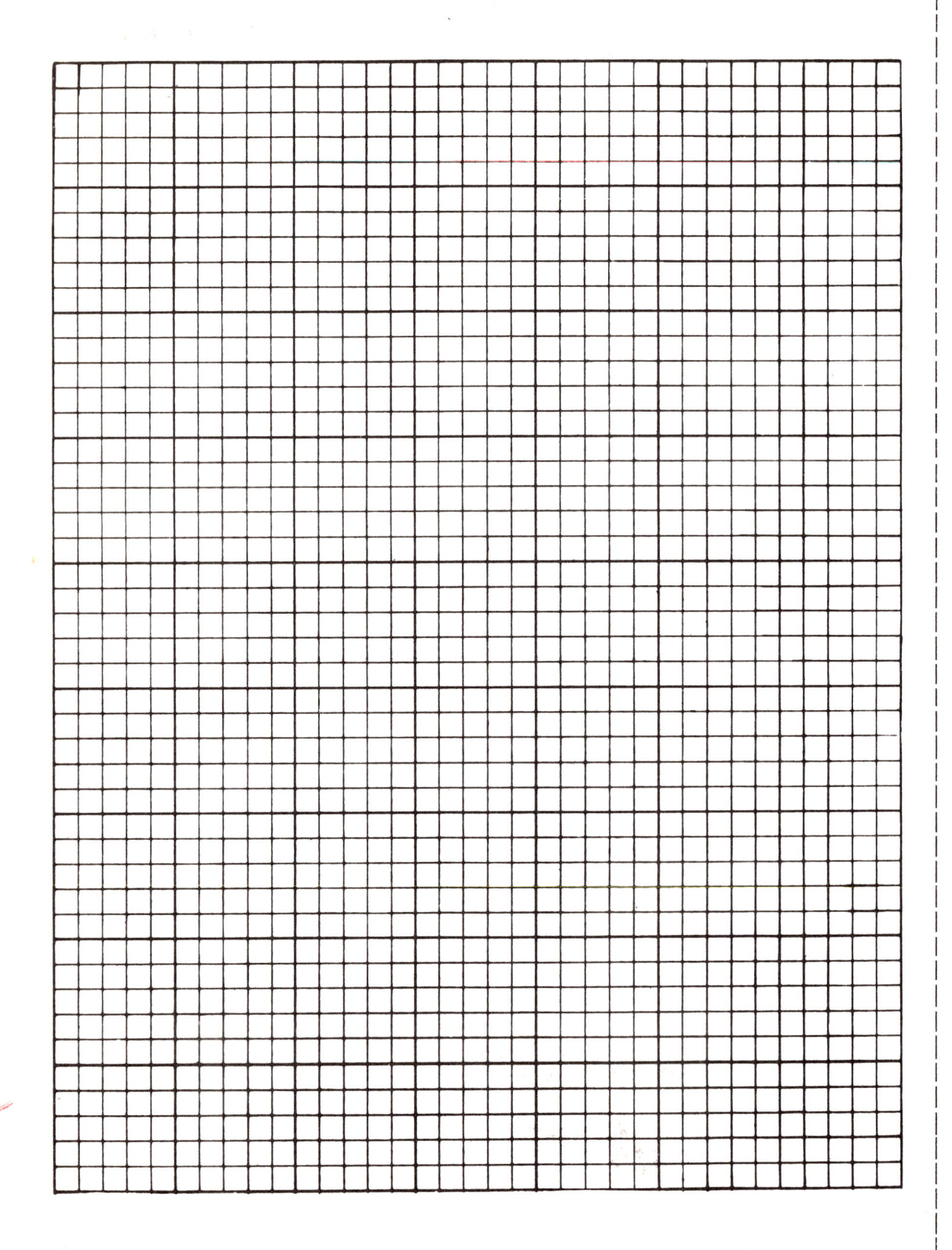